PRAISE ᴦᴜᴋ
THE FLIGHT OF A WILD DUCK

"*The Flight of a Wild Duck* does a wonderful job describing the journey of a young person growing up in the counterculture and political turmoil of the '60s to become one of the major influencers in Silicon Valley. From activist, jazz musician, and a bit more than a geek, Avram was in the room where much of the technology that affects our lives was created. His own role in that was significant. As someone that has known Avram for over forty years, I found the book honest and engaging. This book will introduce you to a truly fascinating, multifaceted individual worth knowing."

—Steven Mayer, cofounder of Atari

"All that I am today, I owe to the moment when I asked Avram to let me join his group at Intel. Much that we now value—interdisciplinary approach, unusual alliances, inspiring storytelling, diverse teams— Avram practiced more than three decades ago. His book also reveals his uncanny ability to predict future technologies and, significantly, to realize their potential impact on society. It was amazing to witness how Avram could make the most talented people in communication, media, and finance become comfortable with technology. The results changed all of our lives. Avram is a compelling speaker. I am delighted to see that he is also a captivating writer. *The Flight of a Wild Duck* will enable many to benefit from Avram's experience and wisdom."

—Lakshmi Pratury, founder and CEO of INK

"Avram Miller has been present at the creation of several seminal events in the computer and communications industries. This wonderful memoir brings his significant contributions to life, with the basic concepts, the underlying trends, the many personalities (Jobs, Gates, Grove, Olsen, Vadasz, and Milken, just to name a few), and the benefits and frustrations of life inside major technological corporations. Find a comfortable chair, grab a glass of something refreshing, and prepare to experience a series of adventures with a wild and highly entertaining 'duck.'"

—Henry Chesbrough, faculty director of the Garwood Center for Corporate Innovation at UC Berkeley, and Maire Tecnimont Professor of Open Innovation at Luiss University in Rome, Italy

"If you want to know how Avram Miller became one of Silicon Valley's most out-of-the-box thinkers at the dawn of the PC age, read his wonderful new memoir. Unlike the current 'Internet Barons,' he lived a whole life before he touched a computer."

—Jonathan Taplin, author of *Move Fast and Break Things* and *The Magic Years*, and director emeritus at USC Annenberg Innovation Lab

"Avram cofounded Intel Capital, bringing the chip giant into businesses that the founders never imagined. He used Intel's money to literally create home broadband Internet. The importance of broadband was obvious to Avram, if a mystery to others. It drove demand for home computers, and it changed both the Internet and the world. The best part about this book is its modesty. It's the story of an inquiring mind finding its way in the world from very modest beginnings and almost no expectations. Avram's life story is about choices. All paths were possible because of his ability and his creativity."

—Robert X. Cringely, host of the PBS TV series *Triumph of the Nerds*, and author of *Accidental Empires*

"It is likely that you are reading this on a device connected to internet. How did we get to this place where powerful personal computers and high-speed networks impact our every day? *The Flight of a Wild Duck* is a unique story of a man that not only had that vision but made it happen by untangling the key technical and business issues. This included navigating the mine fields of corporate politics and stagnant boardroom thinking. We learn how Avram, without any formal scientific or engineering background, was able to deploy his ever-inquiring mind, acute intelligence, and self-taught and disciplined analytical skills to enjoy an exceptionally unusual career, often uninhibited by convention."

—Stephen Maine, former VP of business development at General Instruments

"The vectors that his book spins—from his formative years as an activist, sailing around the world, the various 'loves' in his life—are all a delight. He weaves a story that has so many layers that, if I didn't know him, I'd be asking if this was a work of fiction. This book is sweeping in so many ways. It made me smile, laugh in several places, and, as he details the Intel years, gasp and shake my head."

—Hilton Barbour, VP of marketing at Kognitiv Corporation

"I became general counsel for Intel Capital not long after it was founded by Les Vadasz and Avram Miller. As I worked closely with Avram for many years, he never failed to impress me with his creativity, passion, vision, humanity, and courage. Avram was a great role model in being effective within a corporate setting while being true to oneself. I often wondered how Avram found his path through what must have been for him an alien environment, Intel. Now with *The Flight of a Wild Duck*, I have some of the answers to those questions. This isn't just a book about technology or just for technologists. It's the story of a man who follows his own authentic path to success."

—Suzan Miller (SAM), former general counsel of Intel Capital (1995–2002)

"Much is written about and attributed to Gates, Jobs, Grove, and a handful of others. But these entrepreneurs and visionaries would never have made it to the world stage without dozens of smart, driven, creative colleagues who made personal computing, ubiquitous broadband, and the Internet a reality. These people lived predominantly in the future—inventing the future—creating products and an industry that forever changed the world. Avram Miller was one of these. While

I had the opportunity to collaborate with Avram, I did not know how unusual and inspiring his story was until I read _The Flight of a Wild Duck_."

—Bob Stearns, former SVP of business development, chief technology and strategy officer, Compaq Computer (1992–1998)

"While _The Flight of a Wild Duck_ describes events in Avram's life that occurred years before I met him, I can attest that he has continued on the path of making a real difference to people and companies in his later years. Avram played an important role in the creation of my company, Sommetrics, which has developed a breakthrough in the treatment of obstructive sleep apnea. Avram understands what good science means and what it takes to achieve it. I've known Avram primarily as an entrepreneur and scientific thinker. The book depicts a fuller view of a creative and inquisitive person with enormous resilience. I wish our paths had crossed sooner—it would have made an even greater impact on my life."

—Richard Rose, MD, founder and CEO of Sommetrics

THE FLIGHT OF A WILD DUCK

THE FLIGHT OF A WILD DUCK

AN IMPROBABLE JOURNEY THROUGH LIFE AND TECHNOLOGY

AVRAM MILLER

WIDER
VISION

Published by WiderVision Media, Tel Aviv, Israel
wildduckflight.com

Image credits: Cover and interior images provided by Avram Miller

ISBN (hardcover): 978-1-7372876-2-9
ISBN (paperback): 978-1-7372876-1-2
ISBN (e-book): 978-1-7372876-0-5

First edition

For Dafna (1976–2019),
always my little girl and forever in my heart

You always needed a wild duck like Avram, a nonlinear thinker that stirred up the others. I always try to have an Avram on my team. Always. They disrupt and create.

—Andy Grove, 1999, as told to Renée James

CONTENTS

Part IV: Flying Solo (1996–2002)
Surviving prostate cancer and leaving Intel to be on my own.

FOREWORD

Though our careers in Silicon Valley overlapped by a dozen years, I didn't meet Avram Miller until shortly after his retirement from Intel in 1999. We were neighbors in the rural Wine Country of Sonoma County, sharing the struggle of using long-distance Wi-Fi as the only alternative to crappy telephone dial-up Internet. We literally got to know each other on garage roofs aiming dish antennas at distant ridgetops.

Over the subsequent 20 years I came to know Avram, his children and grandchildren, his wife and ex-wife, even Avram's father, who died a bit more than a year ago. I know most of the people mentioned in this book and I believe I know Avram well enough to love the guy.

Avram is known for his intellect and specifically for his rare habit of asking the question *Why?* Why do we do things the way we do rather than some other way? In Silicon Valley the operant question for most people is *How?* not *Why?* But why is more powerful than how. How is by nature evolutionary while why is *revolutionary*. The other big proponent of asking why in Silicon Valley was Apple co-founder Steve Jobs, which pretty much makes the point of how important why can be.

Dave House, one of Intel's most senior executives—a noted asker of How? who worked with Avram at Intel before moving on to become CEO of Bay Networks, an important networking company—once told me at a Computer History Museum event that Avram was crazy (Hows tend to view Whys that way) "but also the smartest man I ever met."

Avram Miller co-founded Intel Capital and, by doing so, brought the chip giant into businesses that Intel founders Bob Noyce and Gordon Moore would never have considered on their own. Avram used

Intel money to literally create home broadband Internet, even though Intel was not intrinsically a networking company. But the *why* for Intel and broadband was obvious to Avram if a mystery to those others. Home broadband drove demand for faster and more powerful processors in home computers, directly driving Intel's growth and changing both the Internet and the world in the process. Avram Miller did that.

The best part about this book you are about to read is its modesty. It's the story of an inquiring mind finding its way in the world from very modest beginnings and almost no expectations. Avram's life story is one of choices. Did he want to become a rabbi or a jazz piano player? Did he want to be a neuroscientist or a computer pioneer? All paths were open for Avram because of his ability *and his creativity.* The world is a better place for the choices he made.

Young Avram once rented the space under a baby grand piano for a place to sleep. That's a modest beginning and yet charming, like Avram, gaining him not only a place to live but also an instrument to play.

Imagine how far he might have gone had Avram been able to afford to live under a *concert grand?*

> —Robert X. Cringely, host of the PBS TV series *Triumph of the Nerds*, and author of *Accidental Empires: How the Boys of Silicon Valley Make Their Millions, Battle Foreign Competition, and Still Can't Get a Date*

PREFACE

One of the difficulties I faced writing this book was the long period of time covered. My story begins in 1945, the year I was born. World War II was still being fought, Hitler was alive, as was a fifteen-year-old girl named Anne Frank, and the atomic bomb had yet to be dropped. There were fewer than ten thousand television sets in homes and fewer than 40 percent of households had telephones. The first computer, the ENIAC, was not yet in operation. My parents, just children themselves, could not have imagined how technology would change the world on that foggy day in San Francisco when I took my first breath.

This is not a book just about my career. It is about my improbable journey overcoming challenges, from illness to health, from school failure to scientist and professor, from dishwasher and pizza maker to an Intel vice president and co-founder of its venture capital group. I never experienced these challenges as obstacles. Rather, I was driven by curiosity, creativity, intuition, and imagination.

Along the way, I have been fortunate to know incredible people—poets and musicians, scientists and inventors, entertainers and movie makers, founders and CEOs, drug addicts and billionaires. I was comfortable with all of them because I could always make them laugh and I could learn from them. While the book considers those relationships, it also explores the evolution of the critical technologies that ultimately created a new medium that changed how we work, play, and learn, told by someone who played a significant role in its creation.

As I entered my seventies, I realized that the stories, history, and insights I have gathered would go with me if I did not make this effort. It has taken nearly three years and a lot more work than I imagined. Much of that time was spent in research collecting thousands of documents, including articles, internal memorandums, and presentations from various companies. I interviewed more than seventy individuals. I had no idea how much I would learn, not only from the significant research I undertook but even more from the understanding gained by looking at my life's story in reverse. In the process, I not only had the opportunity to appreciate my successes but, importantly, my failures.

Because the book primarily spans the period from 1967 to 2002, one of the challenges I faced was taking readers through this journey knowing that many of them would have very different memories and experiences. Some may even have difficulty imagining a time when it was not possible to access to the Internet from the device in your pocket.

There are things this book will offer in addition to documenting many significant events in technology. It highlights the importance of intuition and creativity and how they can help in achieving objectives, but it doesn't try to sideline the role that luck plays in success. For those who are successful, I hope it will make them a bit humbler. For those who have not found success, perhaps my words will be the inspiration to try again. For those who have been ill, the stories of my illness may offer strength. For those who don't fit in, maybe they will gain an appreciation of their uniqueness. Finally, I hope to leave all my readers with a few laughs.

Avram Miller
Tel Aviv, Israel
2021

PART I

TAKING FLIGHT (1945–1979)

*A sick but gifted child becomes a
seaman, hippie, and scientist.*

The Jewish Exodus to L.A., by Jonathan Kirsch
Ray Bradbury on California's Best Bookstore
Daddy Dearest: The William Saroyan No One Knew

SCOOP! the greatest ice cream

California

REVENGE OF THE NERDS

Remember me? I'm
one of those guys
everybody laughed
at in high school.
Well, today I design
computers and I'm
worth millions. So
who's laughing now?

By Paul Ciotti

1. BECOMING AVRAM MILLER

It's terrifying when you can't breathe. Panic only makes it worse. Whenever I was gasping for breath, my father would turn away from me. He may have felt guilty for passing along his chronic asthma. My mother would weep with fear, which only made me more afraid. When an attack came, I was often taken to Mount Zion hospital in San Francisco, sometimes in an ambulance. I can clearly remember the oxygen, the shots of adrenaline, and the sweet nurses who would care for me.

My parents fought constantly, screaming and hitting each other. Fear of their rage—fear that I would be hurt—often triggered my asthma attacks. Off to the hospital I would go. Eventually, the doctors recommended I be admitted to a convalescent facility. Most likely, they wanted me apart from my parents. And so, at age seven, I was sent away for almost a year to the Stanford Children's Convalescent Home in Palo Alto, California, a facility for treating children who were recovering from serious illnesses.

Stanford Children's Convalescent Home

My school uniform was a bathrobe. I was alone; my parents hardly ever visited me. But it was there, in probably the most critical period of my life, and in that unlikely place, that I developed the imagination, intuition, and humor that equipped me for my improbable life's journey.

As I stared at the ceiling of my room, I began to consider the universe and my place in it. While I was aware of the weakness of my body, I was also aware of the strength of my mind because people constantly told me how smart I was. Curiosity dominated my thoughts. Even though I was ill, I felt that the world was a fantastic place. I was desperate to live because I was desperate to discover and create.

While I believed in God, I did not blame him for my sickness, nor, frankly, did I imagine him even aware of me or my plight. Putting together an imperfect combination of discovery, laughter, sickness, and fear, I took it upon myself to overcome the pneumonia and chronic asthma that kept landing me in the hospital. I came to believe that if I could improve the universe, God would make sure I continued to live. If I failed, I would die. So I began to think that my survival would depend on my actions and contributions. It was both a burden and a calling.

The principles upon which I would base my life began to form. I accepted my illness as part of myself, but I didn't believe that it was my destiny to be killed by it but that I could overcome what was happening to me. Alone in the convalescent home, I spent hours inside my mind, repainting, rearranging, reimagining my room. I played with its dimensions; I rotated the ceiling until it became one of the walls and one of the walls became the ceiling. I changed the walls' colors. I discovered my intuition and imagination—and learned to trust them.

I did my best to make the nurses laugh, knowing they would take better care of me if they liked me. Humor was survival then, and it was a skill that would become very important later: I could have everyone laughing and smiling within a few minutes of entering a corporate boardroom.

As for many kids, Superman was my hero, but I think I admired Clark Kent even more: nerdy on the outside, superhero on the inside. My other hero was Einstein. I liked that he was a misfit and rebel. Like so many in the 1950s, I stood in awe of his genius. My mother had introduced him to me as a figure. When I struggled to tie my shoes, she told me that Einstein could not tie his shoes, either. I was certain she had just made that up, but it helped me feel better about myself. One only needs to look at my hair now to know that Einstein is still my role model. I smile whenever I travel across the world and hear people calling out "Einstein" when they see me.

Some forty years later, I returned to the Stanford Children's Convalescent Home, now called the Lucile Packard Children's Hospital. The memories flooded back as I stood near the one place I had felt safe as a child. I was there because Steven Spielberg had invited me to participate in an event for his Starlight Children's Foundation for chronically and terminally ill children. I had gotten to know Spielberg and liked him very much. We are about the same age and could have been brothers. We both grew up in Jewish families. Of all the people in Hollywood that I've met, he is my favorite because he has such a good heart. I was especially touched when he donated the money he made from the film *Schindler's List* to establish the Shoah Foundation, an organization set up to record the stories of the survivors of the Holocaust in which six million Jews were killed by the Nazis during World War II. My son, Asher, would come to work there.

The building that housed the actual convalescent home, and is now a Ronald McDonald House, was close to the Packard Children's Hospital. What would I have felt if I had returned to that building, to the room where I'd spent so much time living in my imagination?

As I spoke at the event at the children's hospital, I felt great emotional pain as I looked out at the many sick children. I talked with a few of them afterward but did not share my experiences as a sick child, perhaps because I was surprised by the intensity of my emotion. I realized that so much of the pain from my early childhood had never left me.

ARNIE GOLDFINGER

I was born Arnold (Arnie) Stephen Goldfinger at 5:59 a.m. on January 27, 1945, at Children's Hospital in San Francisco. I would later learn that at that moment, in occupied Poland, Auschwitz was liberated. On my birthday, I always take some time to think of those who survived that horrible place and those who did not.

While many people say they cannot remember their childhood, I remember much of mine. I can still see the furniture in our home and recall the meals that we ate, the jokes we told, the parties my parents threw. I remember my toys. I remember the birth of my sister Marlene when I was five years old. I remember my parents fighting. Most of all, I remember the terror of not being able to breathe and the fear of dying.

When I was just three years old, in perhaps my earliest memory, my mother's father, Papa Mark, drove me to an area called Playland at the Beach. We sat in his car and watched the waves. Papa Mark was very kind and affectionate to me, something I didn't get from my own father, and I loved Papa Mark very much. Then, at just fifty years old, he took his own life. My father found him, barely alive, in a coma in his car. It was parked close to the liquor store Papa Mark owned on Fillmore near Post Street, where my father also worked. My grandfather had taken an overdose of pain pills. I eventually learned from my mother about his struggles with gambling, women, and a bad marriage. This would not be my last experience with suicide in my family.

Driving with my mother and father to Mount Zion hospital, she told me that Papa Mark was dead. Somehow, I must have understood what that meant, even at that young age, because I told myself that I had to keep my grandfather's memory alive. I chose a day not too long before when he and I had been to the beach. Every night, I worked to recall vivid details of that day. More than seventy years later, I can still see him in my mind. Every day I look at the framed photo of Papa Mark in my home office.

When I was just four years old, we lived in a duplex building on Eighteenth and California in the Richmond District of San Francisco. I still remember everything about that flat. My mother, father, sister, and I lived on one floor, and my mother's uncle and his family lived in the flat below us, while my great-grandmother Bessie, who owned the building, lived in a small attic room. One day, my mother slashed her wrist in the bathroom. When she came out, blood was everywhere. I remember being alone, although my sister Bev, who was about two years old, must have been there, too. I ran to Bessie, who called an ambulance, which took my mother away. My mother tried to end her life many more times. Sadly, she ultimately succeeded, as did my youngest sister, Marlene. They and my grandfather probably suffered from bipolar disorder, which at that time we could not diagnose or name.

The one thing I always have to remember when I reflect on my family and my childhood is how young everybody was. When I was born in 1945, my father's parents, Bill and Frances Goldfinger, were forty-six. My mother's father, Papa Mark, was forty-eight; his wife, Della, and my mom's mother, became my grandmother at just thirty-eight. Now in my midseventies, I try to imagine them at that age, younger than my own children are now.

My father had one brother, eleven years younger, probably from an accidental pregnancy. I don't think his mother really liked children. She certainly did not like me. My mother was an only child. While I had a vast extended family, I had few close cousins and only one uncle.

We were a family of shopkeepers and tailors, not professionals. My father's parents had a successful camera store in downtown San Francisco. Papa Mark owned a liquor store, and my father, a cigar store. After his marriage to my mother ended, he went to college, eventually got an MBA, and became a successful bank executive. He was the first in my family to get an advanced degree. My mother remarried, to a man named Joel Miller, when I was about ten years old. Joel, at twenty-six, was only sixteen years older than me. They opened a series of donut stores. The first one was on Haight-Ashbury, a location that would ultimately play a big role in my life.

As a young child, I grew accustomed to hanging out at one of the stores belonging to someone in the family. Once I got a bit older, I began washing dishes, waiting tables, and even did some short-order cooking at my mother's places. I supposed that my destiny was to own a retail business like a coffee shop, maybe even a restaurant. I never thought about going to university. As it was, I barely got a high school degree. For both my sisters and me, there were never thoughts of higher education.

Working in a retail business did teach me about money. Cash flow was real, tangible. Removing money from the cash register, my mother would separate cash into different envelopes meant to cover various bills. She explained that it was important to take in more money than would have to be paid out, a lesson many Silicon Valley entrepreneurs have yet to learn.

My sister Beverly was born two years and two days after me. My mother was twenty; she had just turned eighteen when I was born. Three years after Bev, my sister Marlene came into the world. While Bev was always strong and independent, Marlene was not. Marlene's many illnesses included depression. I was very close to Beverly, but I was not close to Marlene. Now I wish that I had been more present in her short life.

Joel and my mother worked incredibly hard. Joel woke up every day around three a.m. to go to one of the donut stores, where he would make all the donuts for all the locations. My mother left around seven a.m. and worked into the early evening. Bev and I would sometimes take a bus to Haight Street so we could see our mother. Often, just the two of us, still young children, would have dinner in the Chinese restaurant across the street from the donut shop. This precise location, Haight and Ashbury, later became the center of the hippie movement.

We had several housekeepers and babysitters. Mimi, a large creole lady from New Orleans, loved us as if we were her own children. She sometimes took us to the movies or to her home on Jessie Street. She put hot sauce on everything and taught me to love spicy food and jazz. Another housekeeper, an older British woman, read us poetry.

Marlene stayed near my mother her whole life, always working in her stores, although she eventually married and then divorced. At twenty-seven, Marlene took her own life. My mother, devastatingly, found her hanging in her apartment. My mother was just fifty-one, and she never got over it. Eighteen years later, she ended her own life with a bullet to her brain.

That day, December 10, 1992, I was in Pennsylvania meeting with General Instruments. Intel was working with them to develop TV set-top boxes and cable modems. When I landed that afternoon in San Francisco, my assistant called me to say that I should phone my step-father as soon as I got home. It was an odd message, and a foreboding one. I knew in my heart what I would hear. My stepfather told me that my mother had shot herself in the head that morning. While I under-stand my mother taking her life, as it was clearly her goal for so many years, using a gun seemed like such an angry way to end her life. I could not help but imagine Joel finding my mother like that. She left me a letter in which she said I would be better off with her gone. In her last moments, she was telling me that she was doing this for me. Perhaps she thought that, but her death left a great emptiness in my heart. In her letter, she asked me to look after Joel, which I did, of course.

While I do not believe that there was anything I could have done, the death of my mother continues to weigh heavily on me. Maybe we can all say this, but my mother made me who I am. I doubt that I could have survived my childhood without the confidence, courage, and unconditional love she gave me.

CHILDREN THEMSELVES

My mother had a miserable childhood. Desperate to leave home, she got engaged to my father on her seventeenth birthday, October 16, 1943. They married six months later and conceived me just a few weeks after that. She became a mother at barely eighteen, a child herself. When I left home at eighteen, she was just thirty-six years old.

She was a complicated woman and made many mistakes in her life. In my early adulthood, I frankly thought she was kind of crazy and unstable. Once I grew older and had my own children, I came to appreciate how difficult her life was: three children by the age of twenty-three and a single mom at twenty-six. Still, she gave me an enormous burden; she told me that she was only willing to live because she loved me so much. I took this to mean that we were in some kind of life-and-death pact. I had to stay alive to love her back, or she would kill herself. Perhaps my mother even needed me to be sick to have purpose in her life.

She was, though, completely nonjudgmental. I knew she would love me no matter what I did. Growing up, I always knew that, as long as she stayed alive, my mother would be there for me. This gave me strength, and I tried to be there for her, too.

My mother attempted to take her own life many times over the years. She would call me and tell me she was going to kill herself. I would then talk her down. These calls could last for hours. My sister Beverly also experienced this regularly. These calls only ended when she finally succeeded.

Like my mother, my father had an unhappy childhood, although of a very different kind. Born in 1923, his asthma kept him out of the military, so he did not serve in World War II. Had he not been asthmatic, it occurs to me now, he would have served in the war, risking injury or death. In some strange way, then, his illness may have created my life.

His first job was stringing cable for the phone company. At the same time, coincidentally, my mother worked for the telegraph company. She moved telegrams from one location at the telegraph office to another, on roller skates, as did all the young women who had that task. At sixteen, my mom started dating my father. On her seventeenth birthday, they got engaged. He was twenty years old.

They were so different from each other that it is hard to understand why they got together, though they were certainly united by a shared desire to get away from their parents. My father was cold, unaffectionate, and judgmental, while my mother was warm, affectionate, and generous. I never asked my mother, but after her death, I did ask my grandmother. Della said that my mother had severe "female issues"; to fix them, her doctor said, she needed to get married and have a child right away. While she had three children, she still ended up having a full hysterectomy at twenty-eight. I did ask my father why he married my mother, and he said it was because she was so sexy. I think there was a lot more to it than that. Though my parents fought all the time, I believe they actually did love each other. My mother kept my father's letters to her from when they were dating. They're with me now, and I have read my father's expressions of love. Many years after they divorced, she still sent him notes on his birthday.

Fillmore District

When my parents were dating, Fillmore's residents were primarily African American. Earlier, they had been mainly Jewish and Japanese. Although the Jews had moved out, a synagogue, Beth Israel, remained, and my parents were married there.

World War II had changed the Fillmore District dramatically. While most of the Jews had already left, the Japanese were forcibly moved to internment camps. African Americans migrated from the south, moved in to work in the defense industry. But the Jews kept most of their businesses in the Fillmore. My mother's uncles were partners in a liquor store at Fillmore and Eddy, near Papa Mark's liquor store. My grandfather gambled, and he also ran a betting shop, where my father worked for a while. After Papa Mark's death, my father opened a cigar store across the street, named Honest John, and kept a back room where he had a small "casino." My dad hated his work, or at least he told me so, but he loved the money. I used to hide on the floor in the back seat of his car when he left for work. I wanted to be with him and get out of the house, but he was always working. Plus, I found the Fillmore District exciting. My dad would act surprised when I got up after he parked, but he must have known I was there.

Just eight years old, I spent the summer of 1953 at my grandmother Della's home in Las Vegas, where Beverly and I would spend several summers. I liked Las Vegas. There was sun unlike the fog of the summer in San Francisco. Della ran the gift shop at the Sahara hotel. Her second husband, Jack, was a blackjack dealer. Della took me with her to work, where I hung out at the hotel's pool. Some of the performers and showgirls watched out for me. Betty Grable, the actress, was my babysitter. Jack and Della took me to see Vegas shows, where Jack taught me to tip. By tipping big, he got us the best tables in the house. Later, I would joke when getting on an airplane that I should tip the captain to get a seat up front in the plane.

That summer, my father left my mother. She called me in Las Vegas with the news. When I returned to our home in San Francisco, he was no longer living there. However, his clothes were still in his closet. I remember taking out a pair of his pants and a shirt and laying them next to me on my bed so I could imagine that he was there with me, but he never returned.

Several months later, my father closed down his cigar store on Fillmore Street and began working for an insurance broker. He studied and got a certification to sell insurance and then got a real estate license. He went to night school while he worked during the day, which earned him a bachelor's and, eventually, a master's degree. He had a very successful career in the banking industry running commercial real estate operations for several major banks. He was very proud of his accomplishments. Sadly, he was never able to acknowledge any of mine. For more than twenty years, he had told me that I was a failure. It hurt me to my core. So later when I became successful, I longed for him to tell me I had turned out alright. But those words never came out of his lips, and now he is gone.

In 1957, four years after he left my mother, he remarried. His second wife, Eleanor, was a schoolteacher from a very proper Jewish family. She was the opposite of my mother, judgmental and controlling, which made Bev and me uncomfortable. Once my father married Eleanor, he put a lot of pressure on us to live with them. We spent most Sundays at their home. But my sister and I could not stand the idea of living there. They were so critical, and our father kept trashing our mother in front of us.

My father, convinced that my mother was crazy, felt that Bev and I would end up badly if we continued to live with her. Sometimes I thought he might be right. He did not care what happened to Marlene, my younger sister, who had resulted from an affair. Marlene was eventually adopted by my mother's second husband, Joel, and she changed her last name to Miller. In Marlene's early twenties, my mother told her about the affair and put her in touch with her biological father, who had been a neighbor. I remember him well, as he came to our home regularly with his wife to play cards with my father, my mother, and other friends. Marlene wanted to meet her biological father, according to correspondence I have between them, but while he was kind to her in his writing, he refused. Soon after, Marlene took her own life. Now that I have lost my own daughter, I know full well the emptiness my mother must have felt.

GIFTED CHILD

In the autumn of 1952, I left the convalescent home and returned to San Francisco. I vividly remember going to the soda fountain across the street and ordering a root beer float. The soda fountain had installed a jukebox while I was away. I don't think we ever listened to music in my home, so the jukebox was fascinating to me. It was my first experience with music. I remember it played many different types. I would often sit at the counter to listen to music.

I was enrolled in the Francis Scott Key Elementary School about four blocks from our home. Although I was not yet eight years old, I was put in the third grade. Trouble started right away. I could not relate to the other students, especially the boys, who seemed rough to me. I was a very skinny and sickly kid. I argued a lot with the other kids, and many times, I got beaten up on the playground. Two teachers took an interest in me, and it is hard to imagine what would have happened to me if they had not come to my rescue. It was decided that instead of taking the regular classes, I would sit and read in the school library. They would assign me books to read, but generally they let me follow my own path. This was a decision that would have major consequences for my development.

I began to read from an early age and was very comfortable spending my days in the arms of books. I was especially interested in science, history, and poetry. One of the most important books I read at that time was H. G. Wells's *A Short History of the World*, which I must have read more than ten times. It taught me to analyze all the events happening around me. I also discovered the work of Margaret Mead, the well-known and influential anthropologist. I loved reading about the Samoans and other indigenous cultures. Mead's work made me feel more comfortable with being different. I remember reading Kipling's poems, and I can still recite parts of "Mandalay."

Every Saturday morning, I listened to a science program on the radio. I was especially interested in nuclear energy, which was a common topic in the culture at that time. Discussions often focused on the possibility of nuclear war between the Soviet Union and the United States. All over my school, signs showed us the locations of shelters. We were taught how to "duck and cover" under our desks, as if that would help us survive an atomic blast.

I had a chemistry set and a microscope which with I experimented and used electricity to try to encourage chemical reactions. This, unfortunately, involved sticking bare wires into a 110-volt power outlet. Good thing I was not born in Europe, or the 220 volts might have killed me. A few times, I made small bombs that went off, setting little fires in my home. I must have been some strange little guy.

— being a gifted child is not such a great gift

Soon I became a subject of a study myself by the state Department of Education, because my IQ was apparently off the charts. I was classified as a "gifted child," but at that point in my life, it did not seem like such a wonderful gift. It only brought me problems, and it kept me isolated from other children my age.

The only class I remember at Francis Scott Key Elementary was math. It was awful. I could solve the problems correctly, but my method was different, especially for multiplication and division. We were taught to multiply one digit at a time, and then carry and add the most significant digit if the result was more than one digit. I no longer remember how exactly I did multiplication then, but I know that it involved complete numbers taken one at a time. My teachers were not amused.

This inability to explain my answers haunted me throughout my life; I often identified the right or best answer in an unusual way, and I often did not even know myself how I got there. As a child, I could look at a page of numbers to add, and I would know the answer without any idea of how I got there. Later in life, I could look at financial statements and immediately recognize any errors. This ability faded with age, as did many other abilities. My mother had a similar capacity, I suspect, as she could complete the books of her business in her mind alone.

After two years at Francis Scott Key, we moved to the suburb of South San Francisco. Then, a year later, we moved back into the city, and I attended Herbert Hoover Middle School in the Sunset District.

The principal was Mr. Zimmerman, the same teacher who had helped me at Francis Scott Key. By that time, I was pretty much failing all of my classes, and I was eventually held back one semester. It would have been easy to conclude that I would not amount to much, as my own father suggested to me many times. Yet I was teaching myself, well beyond my fellow students. I just could not stand to be kept in the classroom.

HOW DOES MUSIC WORK

Then, at the age of fifteen, I discovered music and a more bohemian world. I joined the choir class at Lincoln High School with the hope that this would be at least one class I could pass. I knew nothing about music, and I had very little interest. By that time, my voice had changed, so I sang bass. I began to make friends with classmates, especially Richard and Mike M. Sixty years later, I am still in touch with both of them, especially Richard.

Richard lived just a few blocks away, and we often hung out at his house after choir. Like me, he read a great deal and was interested in science, philosophy, history, art, and most importantly, music. Finally, I had a real friend with whom I could share so many of my interests. We were very different, though. I was extroverted, with a big personality and a lot of drive. Richard was an introvert. Richard tells me now that he was never surprised by my later success.

Richard composed music and aspired to do so professionally. He had learned to play piano just a year before we met. Though he did not play that well yet, he could compose. He still shares pieces he has written with me. By contrast, Mike M. was a great pianist. Listening to him play Chopin études, Beethoven sonatas, and Brahms intermezzi, I was amazed and dazzled by his talent. Trying hard as I was just to stay in the choir, I did not yet think of learning to play an instrument myself.

Richard knew much about music history and introduced me to Western music: chants and modes; Renaissance music; and music of the Baroque, Classical, and Romantic periods; as well as, later, the Impressionists; and eventually, the contemporary atonal composers.

One day, the scientist in me led me to ask him a curious question: "How does music work?" And so Richard began to teach me music theory. He gave me textbooks to read, the most important of which was by Paul Hindemith. I started doing the exercises in that book. The great thing about singing in a choir was that it gave me an excellent way to learn four-part harmony. Learning to hear the inner voices profoundly fostered my own musical development.

We had a small upright piano at home because my sister Beverly had wanted to take piano lessons. Though she soon lost interest, I taught myself musical notation. I sat down one day at the piano and started to write a piece of music. I can still remember some of it. The high school parent-teacher association had a scholarship program for various topics, including musical composition. I applied and was awarded enough money to study composition and orchestration after school at the San Francisco Conservatory of Music. I studied piano, and I picked up a few other instruments, including the trumpet and clarinet. My teachers at the conservatory told me that I'd started too late to ever play an instrument well, which I accepted as true for many years. While they may have believed this to be the case for classical music, I have now studied jazz piano almost my entire life. I certainly do not believe that one can only play jazz piano well if one has started as a young child.

Playing the piano and composing kept me sane and grounded throughout my life. Jazz piano in particular is perfect for me, as it lets me use my creativity and quick mind. Jazz is highly intellectual and mathematical, but it also requires a sense of time. Most importantly, for me, is that playing music forces me to exist in the moment, something with which I have always struggled. In that respect, it is a bit like meditation.

Much later, in 1995, Ann Bowers, the widow of Bob Noyce, Intel's founder and the inventor of the integrated circuit, asked me to serve as an advisor to the Noyce Foundation, which was formed after his death to honor his memory. At a dinner one evening for some of the advisors, I found myself sitting at a table with Murray Gell-Mann, the well-known physicist and Nobel Prize winner. The others at the table were mostly mathematicians. As the discussion turned toward mathematics and intelligence, I ventured that jazz musicians, and especially pianists, performed complex mathematics. One of the math people laughed. "What they do is nothing compared to what we do," he insisted. I countered, smiling, "All right—but can you do it in your head and on the beat?"

MY FIRST REAL JOB

I hardly ever attended my high school classes, aside from choir. I showed up to tests, passing some and failing others. I claimed that I was sick all the time. Since the attendance office only checked the first class (homeroom), they did not know I came in for choir.

Later, I was allowed to take a work/study program, which meant that I only had to attend classes in the morning. I worked at a jewelry store in the Stonestown mall four hours a day, six days a week. I picked up all the jewelry pieces left for repair and took them to downtown San Francisco to several vendors, picking up completed items to then take back to the mall. I thought it was funny to be carrying jewelry worth thousands of dollars in a small leather satchel on the streetcar downtown and back. Each round trip took an hour, and I would spend that time reading. I went back and forth between the store and downtown five times a week. On Saturdays, I stayed at the store and polished silverware. I hated that, but otherwise I enjoyed my work.

I did two things with the money: I bought clothes, and I purchased a baby grand Baldwin piano. I very much wanted the piano, which I was able to buy (used) by making a deposit and paying in installments. I kept that piano for more than thirty years before replacing it with a larger Steinway and, later, two other grand pianos. My interest in clothes is perhaps a bit of a mystery, though I did come from a family of tailors. In the '90s, I joked that I was the original metrosexual, and I particularly enjoyed helping women buy clothes. Once, in an interview with *USA Today*, I made a public relations mistake: I said that had I not gone into technology, I would have gone into fashion and designed women's clothing. After that interview came out, many women contacted me asking me to design clothes for them. I think that might have been just an excuse for calling.

By the time I was seventeen, I was failing almost every class, and it had become clear that I would not graduate from Lincoln High. My mother, always resourceful, got the expensive Drew School in San Francisco to agree that if I attended for a year, I could take their tests. If I passed, they would give me a high school diploma. Of course, I hardly ever attended classes there, but after a year, I did take the tests, and I passed. I was not uneducated, merely self-taught. I only wish that I'd had access then to the capabilities to learn that exist now through the Internet.

There I was—a skinny soon-to-be eighteen years. The asthma that had defined my early years was under control. My high school diploma had been not so much earned as purchased by my mother, who had cleverly sent me to a private school that let me test out, and there was no thought of continuing my formal education. My friend Richard was already in his first year of college. I, on the other hand, had no idea what was awaiting me.

2. AN ORIGINAL HIPPIE

Throughout my childhood, my mother was the biggest force in my life. But as I reached adulthood, she offered me no direction. There must have been some anxiety in the family about what would become of me, however, because soon, I was presented a unique opportunity.

GOING TO SEA

My grandmother's second husband, Jack Silverman, was a merchant seaman during the Second World War. Through his friends in management of the Marine Cooks and Stewards Union, he got me a place in their three-month steward training program in Santa Rosa, California. The school was also the retirement home for members of the seaman's union, who were, in effect, our customers. Bound for employment on luxury liners, we were trained to offer first-class service. The good news for us was that the meals were also first class, and we ate the same food that we served. I still have many of the skills I learned as a steward.

President Cleveland in Hong Kong, 1963

One student, older than most, Nate Freeman, had actually been the head waiter at a well-known dinner and comedy club in North Beach called The Purple Onion, which hosted headline acts like Lenny Bruce and Woody Allen. Nate had enrolled because he decided that he wanted to work on ships instead of clubs, and we became good friends. His wife was Japanese, and they would often invite me to spend the weekend at their Bush Street apartment in San Francisco. His wife was chaperone to beautiful twin Japanese girls who were singers and eighteen, like me. I paired off with one of them, but we always went out in groups and were never left alone. As they performed at a Japanese club, we would get together after their last show and go out to an after-hours club in Chinatown to dance.

My "girlfriend" did not speak English, and I was struggling to learn Japanese. Her sister had a Japanese boyfriend. The four of us were once allowed to go out together by ourselves for a picnic. I had a camera and took some photographs. My father's father, Bill Goldfinger, had a camera store, and he would develop my pictures for free. So I gave him my film from our picnic, without thinking about his reaction once he saw them. The extent of my grandfather's prejudice against the Japanese surprised me, and he became distraught and angry. We did not speak again for many years. Though I thought he had forgiven me, I was the only one of his grandchildren to be left out of his will when he died forty-five years later.

Soon after I had graduated from the steward's school, I went to Zim's, my favorite hamburger joint. After I finished my cheeseburger and left, I felt a terrible pain in my back on the left side as I was about to open the door of my car. With no cell phones in those days, I went back into Zim's and called my mother. She said she would ring our family doctor, and I waited for her to call me back on the pay phone with his instructions. The doctor said I should go to his office immediately. After examining me, he sent me right then to Mount Sinai Hospital where I was admitted on the spot.

CAN'T BREATHE AGAIN

I submitted to several examinations, and my mother arrived. The doctors explained that I had a spontaneous pneumothorax. This meant that there were several leaks between my lung and my chest wall, which created air bubbles that eventually collapsed my left lung. Over the next week, the doctors tried many procedures to repair my lung, but they all failed. I was in great peril. Finally, the doctors took the most drastic action: they opened my chest and adhered the back of my left lung to the chest wall so there would be no place for air bubbles to develop. I lost half of the capacity of my left lung and several ribs.

My operation and hospital stay were costly, of course, and my mother asked my father to contribute. He walked into my hospital room and said, "If you think I am paying for any of this, you are crazy." I got sick to my stomach and threw up. I had just eaten a bacon-and-avocado sandwich. I would neither see nor speak to my father for sixteen years. It was also many years before I could eat avocado. Fortunately, I have not spent one night in a hospital since April 1963. A scar on my back is pretty dramatic reminder of that operation. When people ask about it, I often say I got it in a sword fight. It's funny how frequently people believe that.

Three weeks later, and against the advice of my doctors, I took a job as a steward on the SS *President Cleveland* and sailed to Asia. I was still just eighteen years old, and so was the *President Cleveland.* The ship, originally meant as a troopship, was laid down in 1944. With the war ending in 1945, the American President Lines fit it out as a luxury liner. It could carry 579 passengers, and 379 in first class.

As a member of the Marine Cooks and Stewards Union, I had a blue card to submit as an application for jobs aboard the ship. There was a caste system: those who were qualified to interact with passengers had a blue card, while those who did not, such as dishwashers and laundrymen, had a red card. Whoever had waited the longest for a job got it. If you did not get a job in ninety days, your ninety-day wait started over. Because of the operation, my ninety days had almost expired, so I took a less desirable job: steward to the officers of the *President Cleveland.* I cleaned their rooms, made their beds, and kept the shared cabin area clean. Basically, I was their maid. The good part of the job was that when we were in port, the officers would take off for a few days, so I was free as well. Often, in Yokohama, they snuck prostitutes into their rooms, and then dropped them off to the south in Kobe. Since they would not allow me to enter their rooms for that duration, I ended up with even more free time.

I explored Yokohama, Kobe, Hong Kong, and Manila. This part of Asia was very different in 1963 compared to today. After the ravages of the war, Japan was a developing nation. Hong Kong was full of "boat people" sleeping in the streets on bare mattresses. One terrifying night in Manila, I was chased through the streets by robbers with guns. Japan I particularly loved, since I could speak a bit of Japanese and get around. One of my Japanese friends on the ship took me to a bar in Yokohama, where there were "girls." One of them, who was my age, took me upstairs to a room, and there I lost my virginity. Soon after, I returned to the bar, and my friend came down accompanied by two girls. Surprised, I remarked, "I didn't know it was possible to be with two girls at the same time."

On my first night in Hong Kong, I made the mistake of stopping by one of the many bars close to the docks in Kowloon with several of the ship's officers. They all wanted to buy me a drink. I was only eighteen and had minimal experience with alcohol, so I was soon drunk. We decided to visit the island that is Hong Kong proper, which at that time required crossing by ferry. I bought the cheapest ticket, on the lowest level, and stood on a deck just above the waterline. The people near me, all manual laborers dressed in black uniforms, must have thought it strange to see a peculiar, skinny white boy standing in their midst.

Somehow, I slipped and fell into the water. The ferry stopped, and I was pulled back in. I arrived on Hong Kong Island wet but not cold, as the weather was rather warm. I decided to go to the top of Victoria Peak by tram. There I stood looking out at the bay of Hong Kong. I thought about a movie set in Hong Kong I had seen as a child: *Love is a Many-Splendored Thing*. I could not have imagined then that some-day I would sit on the board of a company that owned the Hong Kong telephone company and would have meals with the head of the chief executive of Hong Kong.

The *President Cleveland* itself was a dangerous place. Some of the sailors were pretty scary, and fights were common. Such disputes were resolved with knives and, once, a gun. I heard of sailors thrown over-board during some fights, but I never saw it happen on ours. The sail-ors were divided into three groups. The first, us straight men, lived on the starboard side of the vessel. The port side was mostly gay crew members. My roommate at steward school had been Asian and gay, and we were friends. Many of the other students were gay. But I had no idea about this division until my first evening on the ship, when I decided to take a walk around. On the port side, I saw a large party with what I thought was a lot of women, but I later discovered it was men in drag. The third group, Chinese men, worked as cooks and in the ship's laundry. They did not speak English and kept to themselves, sitting on their haunches and eating out of bowls in the laundry area.

Much of the crew gambled a lot, with the games run by Chinese crew members who spoke English. They paid other people to do their jobs for them, because their real business was running the gambling. Some of the crew would gamble away everything they made. When the *President Cleveland* docked back in San Francisco, the men who ran the gambling ring would be there to collect their gambling debts, since we were paid in cash as we left the ship. While I came from a family of gamblers and card dealers, I never liked gambling myself.

After six months sailing back and forth across the Pacific, I decided that merchant seaman was not to be my profession. But I had made and saved a lot of money. I moved back in with my mother and Joel for a while, then spent some time with my grandmother Della in Daly City. I had some difficulty breathing while I was at sea, so I went to be tested at the Maritime Hospital in San Francisco. They told me that I had early-stage emphysema and that I was not likely to live longer than another ten years. I tried to push thoughts of my mortality out of my mind. I vividly remember coming home after my diagnosis and telling my mother I had about ten years to live. She cried, but we were on familiar ground. I was also smoking two packs of cigarettes a day at that time. I would not give up cigarettes for another eight years, until I was married and became a father. Having ten years to live is long enough to ignore the threat of dying.

FIRST LOVE

At that time, my mother and Joel once again owned donut stores. I started working in one of them, playing piano, and hanging out with Richard. One evening in 1963, Betty, a friend of mine, invited me to join her to celebrate the sixteenth birthday of her friend Holly. I had seen Holly, who I found very attractive, at a party a few weeks earlier, so, of course, I agreed. Once I arrived, it became clear that Holly and I shared a strong mutual attraction. As the evening came to a close, I invited Holly to join me in driving Betty back to her home. After dropping Betty off, we drove to an isolated spot, and Holly and I "made out." I was in love for the first time in my life.

Holly and I became extremely close. We started going steady, as people said at the time. She was an amazing young woman: beautiful, smart, and talented. While Holly was an excellent student in all subjects, including science and math, her love was music. She sang soprano, and I sometimes accompanied her on the piano. Holly's parents were very liberal; however, they grew concerned about how close we had become, and did not like that we were sexually active. Things became more than a bit uncomfortable. They thought we were too young for sex, but actually, we were too young for love.

Holly's father's family was originally from Spain, and she spoke beautiful Spanish. We decided to move to Spain when she finished high school. At nineteen, I left for Barcelona. The idea was that I would spend that year learning Spanish, and then Holly would join me the year after, once she graduated. What we would do there or how we would live was not really thought through. This move was a huge mistake. I could not handle being separated from her. Within two weeks, I returned to San Francisco, somewhat defeated.

When Holly graduated high school in 1965 at eighteen, I was twenty years old. Spain was no longer the plan. Instead, she would attend the Monterey Language Academy and study French. A few weeks after her enrollment, I drove down to Monterey to spend the weekend with her, where she had rented a small house. I was very excited since we had never been together like that. As Holly walked toward me to greet me, I sensed that something had changed. I knew our relationship was over and immediately said as much. She agreed. Evidently, her parents and friends had convinced her that she was too young to be in such a committed relationship. I was devastated. I stayed over that night and drove back to San Francisco early the next day. My heart was broken, and I actually never really got over it. I refused to let Holly out of my life. I told her that I accepted the end of our romantic relationship but that I wanted to stay in her life as long as I lived. And so, I did. I have been there for her throughout everything that took place in her life. I remain very close to her and to her two daughters. In 2013, we celebrated our "fiftieth" anniversary. We visited the places we once frequented, and my old friends Richard and Betty joined us to celebrate.

BLUE UNICORN: WHERE IT ALL BEGAN

In the early 1960s, San Francisco underwent a major cultural shift driven by music and drugs. North Beach was full of small cafes that would host folk singers and poets. Singers like Bob Dylan and Joan Baez were becoming popular. While living at home after my stint as a merchant seaman, I was lucky enough to meet them both. The building next to the synagogue where my family went became the Fillmore Auditorium, developed into a famous music venue by promoter Bill Graham that hosted performances by many rock groups, including the

Grateful Dead, Jefferson Airplane, Jimi Hendrix, the Doors, and Pink Floyd. Most of the Jews in San Francisco eventually moved to the suburbs, and the synagogue followed. At one point, the building next to Beth Israel became the Peoples Temple, hosting the infamous Jones Cult.

One day in 1963, I stumbled onto the Blue Unicorn, located in the nearby Haight-Ashbury district of San Francisco. The hippie movement started, I would say, at the Blue Unicorn, which was a pretty small, long, and narrow cafe. Besides coffee, they served some cheap food like hot dogs. I became a regular. The cafe was owned by Bob Stubbs. His younger brother, Norm, was my age, and we became good friends.

The cafe had a piano, which I would sometimes play. Mostly, I hung out, drank coffee, and had long discussions with the other people there. Holly would join me from time to time, but relatively rarely, since she was still in high school. Some fantastic people would pass through, including the poet Allen Ginsberg. When I was younger, Richard and I would often go to readings at the City Lights Bookstore in North Beach owned by the poet Lawrence Ferlinghetti, who would often work the counter. North Beach was the center of the Beat Generation literary movement in the city. I remember hearing Allen Ginsberg read poetry at the North Beach Catholic Church around 1960, when I was just fifteen years old. Through the Blue Unicorn, I eventually become very friendly with Ginsberg, who was much older than me. We had long conversations, and he invited me to his apartment several times.

Around then, I met Paul Kantner of the band Jefferson Airplane. Holly and I sometimes visited the band's house on Page Street and smoked weed there. At that time, I was living at my grandmother's home in Daly City, where I had my Baldwin baby grand. I put a mattress underneath the piano and slept there, which I somehow thought would inspire me to practice more. Paul visited me there a few times, and I tried to help him arrange some pieces. I did not like rock, which was all around me, as I was personally committed to jazz. Several years later, I had a chance to help out at a few rehearsals of Janis Joplin and Big Brother and the Holding Company. She was a fantastic singer but did not seem like a very nice person. I did like her music. I also went to a few performances at the Fillmore, next door to where my parents got married. It was not my scene, thankfully, as I might not have my hearing now. Though this was a vital period in the development of American music, and many of these people and much of their music became world famous, I was happily oblivious.

Broadway Street in North Beach had several jazz venues, too, including the legendary Jazz Workshop, and that scene interested me far more. While underage, I snuck into the nightclub and heard some of the great jazz musicians play, like Charles Mingus and Miles Davis. Later, Broadway Street would be lined with a bunch of topless clubs, one featuring the legendary Carol Doda. The Jazz Workshop continued there for a long time.

In the Fillmore District, musicians of all genres would gather after they had finished their gigs. I had many musician friends, and I often joined them there. It was grim. Musicians were taking serious amounts of hard drugs. Two of my friends eventually overdosed on heroin. I never took hard drugs, though I did smoke marijuana occasionally. I did not like to play while high. I would joke that my friends wanted to get high and then practice, while I wanted to practice and then get high.

I was curious about psychedelic drugs. When Norm Stubbs asked me one day if I would like to try LSD—which had spurred much discussion in the media, thanks in part to Timothy Leary, who had been forced out of Harvard for his work with the drug—I agreed. Norm suggested that we first spend the evening at the Magic Theatre, located on Divisadero Street in San Francisco. The Magic Theatre took its name from a 1927 novel by Hermann Hesse, *Steppenwolf,* that had become popular in the hippie culture at that time. I had read all of his books in a course at the University of California, and they had influenced me greatly. So the suggestion seemed natural to me, and I believe that they had a dance performance that evening.

Afterward, at Norm's house, I took the LSD, which was contained in sugar cubes. Norm was to be my guide. For an acid trip, you need one. Sometimes, people would try to jump out of windows because they thought they could fly. Often, people would have a "bad trip" and become very paranoid. I sat there in Norm's living room, waiting for my journey to begin. After an hour, the color on the walls, the ceiling, and the furniture dripped down as if it were boiling oil, mingling on the floor in a big swirl of color. Everything else now became white. It was a bit like my experience as a seven-year-old at the convalescent home, where I would change the colors of my room using my mind. From this world of color, I was taken up into the universe and flew around it as if I were a human spaceship. I don't remember much of the next three or four hours. I am sure that I talked with Norm about my experience.

When the sun came up, I wanted to go for a walk. There, I had one of the most profound experiences of my life. The colors I saw were now much more vivid than I had ever experienced, and my sense of three-dimensional space was much more intense. The external world had been altered and was now far more vibrant and lovely. Thankfully, these changes stayed with me. I had never really seen how beautiful the world is. We walked through Golden Gate Park, looking at the grass, the trees, and the flowers. We ended at Holly's house, many miles away. I don't know if I had told her of my plans to take LSD, but I showed up at her home and we talked. I was still very much under the influence of the psychedelic.

Later, as I rode in a car, it turned, and somehow in my imagination, I didn't. I felt and saw myself continuing out of the car and into traffic. Though it lasted just a moment, I was quite frightened. By the end of that day, the effects of LSD were gone. Although, as I said, my perception of color and depth was forever altered, I had no interest in ever retaking it.

After that, I wisely realized that a career in music was not for me, mainly because of all the drugs.

TAKING TO THE STREETS

Around the same time, I participated in several sit-ins associated with the Civil Rights Movement, including ones at the Sheraton Palace Hotel and Auto Row. At the Palace Hotel sit-in, I was arrested. My name at that time was still Arnie Goldfinger, and they booked us at the jail alphabetically. The comedian Dick Gregory was standing behind me in line, because our names both started with G. When I got to the front of the line, the police officer at the desk asked if I had any scars or incisions. I told him about the scars from my lung surgery. He asked if anything had been removed. Unable to help myself, I quipped, "Yes, they removed both lungs." The officer duly wrote that down in my record. Dick Gregory laughed like mad.

I was one of the eighty-one arrested.

My interest in civil rights was influenced by the time I spent on the streets of Fillmore near my father's cigar store. Back then, most white kids knew few Black kids or adults, but I knew many, and I could not understand why there was discrimination. I also knew what had happened to the Jews in Europe and had even personally experienced some antisemitism. I saw on television or read in newspapers about civil rights protests in the southern United States, led by people like Martin Luther King Jr. When protests were finally organized in San Francisco, I joined.

Whenever we protested, we saw men who were clearly undercover police, perhaps even FBI agents. I could see them photographing us as we protested. Pretty sure they were tapping my phone, I used to pick up the phone and joke, "FBI, get me the operator." I made a device that I could connect to the phone line and generate a loud sound that would have hurt someone's ears if they happened to be listening on the other side. I imagined with pleasure the pain I was causing these FBI agents listening in on my phone line. I am not sure they ever did. Without luck so far, I have tried to learn if I have an FBI dossier.

In 1963, I helped register voters for Willie Brown's successful run for the California House of Representatives. Registering voters in the Fillmore District as a young white man was both challenging and dangerous. Brown, an African American, eventually became the speaker of the California State Assembly, a role in which he served for fifteen years before being elected mayor of San Francisco.

Later, in 1966, I became involved in the anti-war movement. I laid down in front of semitrucks from Dow Chemical to stop the delivery of napalm to ships in the port of Redwood City, bound for Vietnam. The napalm bombs were dropped onto enemy villages, burning up everything and everyone, including children. I thought the war was horrible, and I also believed that the government was lying to us. As it turned out, they were.

My chronic asthma made me ineligible for the draft myself, but I coached others on how to avoid it. One of my good friends who had been drafted left the United States rather than enter military service. Like many others who refused to serve, he was a conscientious objector. I think he was not allowed to return to the United States for more than two decades.

OFF TO EUROPE

After Holly broke up with me in 1965, I moved to Paris, where another friend, Mike R., was studying guitar at the Paris Conservatory. I stayed at his apartment and took a few classes at the conservatory myself. With little money left from my days as a merchant seaman, Mike rented me space under his dining room table. He was pretty strange, to say the least, but we both shared an interest in electronics. We built several ingenious devices. We also bypassed his electric meter. The electric company would give us notice when they were coming by to read the meter, which was inside the apartment. In response, we would dismantle the bypass, and then light the apartment with candles. We'd explain that we didn't use electricity because we were so poor.

We learned that a high-quality portable tape recorder had just been released in Germany, and so we traveled there to buy it. What had happened in the war made me apprehensive to go, but Mike convinced me. We stayed at the Bunker Hotel, just across the border from Alsace-Lorraine. The hotel had actually been a bunker during the war. It was just twenty years after the end of the war, and I felt sure that everyone I saw older than thirty-five was a Nazi. Many no doubt were.

Once back in Paris with the recorder, we figured out how to record the sounds that the pay phone made when you deposited coins. In those days, if you made a long-distance call from a pay phone, the operators would ask you to deposit a certain amount of money to speak for three minutes. Each type of coin would make a distinctive sound on the line, which the operator would track and credit. We recorded the sounds when we made a call. From then on, we just used the recordings to reproduce the sounds that would tell the operator that we had deposited coins.

I went to London to visit my mother's uncle, Joe, and his wife, Alice, for two weeks. I had a great time. When I returned to Paris, I decided not to stay. I did not like the Parisians. Though I tried to learn French, Parisians were unhelpful and intimidating. If you spoke French poorly, they assumed you were stupid. Just before turning twenty-one in 1964, I went back to London, where my grandfather Mark's siblings lived, and where I felt much more comfortable.

NO JEWS WANTED

I rented a bedsitter, just one room with a bed, sofa, chair, and desk. When looking for a place to live, I was shocked to see advertisements in newspapers saying, "No Jews wanted." I had not much experience with antisemitism before. This was the beginning of the realization that Jews were still not safe in Europe.

All of us in the building shared the kitchen and bathroom. In addition to me, the building housed five families from Nigeria. My room had a heater that required ten-pence coins to operate. It took me little time to figure out how to get around that so I could have all the heat I needed. The hot water in the kitchen and in the bathroom also required coins. I did not hack those units. I did not cook there, and I took baths at my uncle Joe's apartment. Once each week, I had what I thought of as my "weekly treat." I took the Underground down to Oxford Circus station, where I would buy a Sunday paper. Then, I went into one of the stalls in the men's room that required a coin to use. I could take my time while I did my "business," reading the newspaper. It was lovely not to worry about one of my Nigerian neighbors knocking on the bathroom door.

My days were full of writing poetry and exploring London. I liked to go to Speakers' Corner and listen to people ranting. I sent some of my poems to Allen Ginsburg, but he never wrote back. In doing research for this book, I was shocked to discover that he had kept my poems and they are in a collection of his papers.

My aunt Alice had a nephew my age, and we became friends. Then I became friends with his friends, so I had a bit of a social life.

I generally enjoyed my stay in London, but I remained heartbroken about Holly. With no idea of what I wanted to do with my life, I decided to immigrate to Israel. But first, I wanted to explore more of Europe.

The air ticket I had bought to get to Europe originally allowed for some additional destinations. Leaving London, I went to Brussels, Amsterdam, and Copenhagen.

After Copenhagen, I set off by train to my ultimate destination, Athens. We crossed into East Germany, which was controlled by the Soviet Union, and arrived in East Berlin in the morning. The train to Prague, my next stop, was not until late in the evening. I wandered around East Berlin and discovered that I could effortlessly cross into West Berlin via Checkpoint Charlie. The two parts of Berlin were remarkably different.

Back at the train station in East Berlin, some Russian soldiers hassled me, because they thought my beard was funny. They kept pulling it and teasing me. At twenty-one years old, with long, curly black hair and a big black beard, I was scared. I weighed about 110 pounds. I was so skinny that I looked like a human paintbrush, so I was thrilled to get onto the train for Prague. Eastern Europe was very cheap, and I could convert my money on the black market. Suddenly, I was rich. I stayed at a lovely hotel for about five days.

I left Prague by train for Belgrade and after twenty-four hours arrived in Athens where I planned to spend a few weeks before departing for Israel. But I changed my mind and decided to return to London. Now I realize that had I immigrated to Israel back then in 1966, I would have become a soldier in the Six-Day War in 1967 and perhaps died.

Upon arriving at Heathrow Airport, immigration would not let me enter the United Kingdom. I told them to get in touch with my family there, which they did. Nevertheless, the immigration officers said that they would only let me stay in the country for a couple of days. Two days later, I was on my way back to San Francisco.

While I was traveling in Europe, a friend Aaron Gonzales had written to me about a program at the College of San Mateo, or CSM, where he was studying music. I had taken some courses there in 1964, when I was nineteen years old. The program Aaron had mentioned, the College Readiness Program, was started by a remarkable woman named Jean Wirth. Aaron said that if I returned to the United States, I could take a room in the house he had rented with his wife in East Palo Alto. And so, I did, once I returned to the United States in June 1966. East Palo Alto, at that time, was a predominantly African American community. The town itself is in San Mateo County, on the edge of what is now Silicon Valley.

I applied to work at the College Readiness Program and was accepted. We worked with high school graduates from East Palo Alto who had been accepted to various universities but did not meet the qualifications to enroll. Our mission was to provide them with some of the skills they would need to be successful in university. That role for me was a bit ironic, given that I had never attended university myself. I loved working with the students, who were about four years younger than I was, and with the other staff members. I made a lifelong friend there, Geoff Wells, and the program itself has had many fantastic successes over the years.

SHLOMO CARLEBACH

While I was working at the College Readiness Program, a friend, Rita Vrat, told me about a Hasidic rabbi, Shlomo Carlebach, who wrote and sang songs. Rita asked me if I wanted to meet him. By this time, Shlomo had performed at the 1966 Berkeley Folk Music Festival. One evening, Shlomo was holding "court" at the home of one of his followers. Meeting him was one of the highlights of my life. He was warm and affectionate. As it turns out, he was too affectionate with some of the women who followed him. This was around the time of the Summer of Love, though, and I loved his music. He sang and played guitar, while those gathered would sing along. He introduced us to Hasidic mysticism, including the teaching of the Baal Shem Tov.

I spent a great deal of time with Shlomo and some of his other followers. I even traveled with him on a few occasions. He called me Avramela. I started wearing a kippah and tzitzit, laying on tefillin during prayers. I started keeping kosher and studied Jewish mysticism. Shlomo bought me several gifts, like a wine cup or a kippah, some of which I still have. As I believed I had fewer than ten years to live, I wanted to spend them as a mystic. I decided I would go to rabbinical school in Israel. This would be the second time I planned to go to Israel and did not.

Shlomo around 1966 in San Francisco

But I had picked the trumpet back up, and I wondered how I could play the horn with the lung disease I had been told I had. So, I decided to get tested again. This time, the doctors told me that I didn't have emphysema. Instead, they explained, I must have had some low-level asthma symptoms, which were somehow mistaken for emphysema.

Once I got my new diagnosis and did not expect to die early, I no longer wanted to be a rabbi. By this time, however, I was broke. I took a job as the night manager at a pizza parlor while I searched for a more suitable position during the day. One of my friends from Shlomo's circle, Mel, was a resident at the medical school of the University of California in San Francisco. He worked with a researcher, Joe Kamiya, who was looking for someone to help him design equipment for some experiments he was interested in performing. Mel arranged a meeting.

ARNIE GOLDFINGER NO MORE

After I left the merchant marine and became very active with the civil rights and anti-war movements, my name and photo would often show up in local newspapers. I was even interviewed a few times. My father, Clarence Goldfinger, was incensed; he knew that if his business associates realized his son was an activist hippie, it could harm his career. No one could know that he was my father.

When I was twenty-two, my mother approached me and said that my stepfather, Joel, thought of Beverly and me as his children. He had already adopted my youngest sister, Marlene, many years before, and now, she said, Joel wanted to adopt us. Knowing how manipulative my mother could be, I suspect that she had told Joel that Beverly and I thought of him as our father and wanted him to adopt us. Well into adulthood at this point, I really had no interest in doing this, but I didn't know what to say. So I asked, "What will Clarence think about this?" My mother said that she had already talked with him, and he thought it was a great idea. Not too long after, Joel legally adopted me. Even my birth certificate was changed to show that Joel Emerson Miller, age seventeen upon my birth, was my father. I was already using my Hebrew name, "Avram," from the time I had met Shlomo. And that is how I became Avram Miller.

In 1981, twenty-two years later, Clarence, my biological father, no longer my legal father, contacted me through Beverly. They had been communicating for several years. I lived close to Boston at that time, and Clarence was traveling to Boston on business and asked my sister to find out if I was willing to meet him. I had been furious at him after that day in the hospital in April 1963, but by the time my sister asked me, I realized that my anger had gone. We met, and there I was, talking with my father after more than two decades over dinner at his hotel in downtown Boston.

The last time I had seen him, Clarence was just thirty-eight years old. Now, he was fifty-seven, with gray hair. I was thirty-six, not much younger than my father had been when we had last met. He showed me a letter I had written him from Paris when I was twenty years old. In the letter, I had tried to reconcile with him. Though he had never replied, he kept the letter, which I had completely forgotten. I asked him how he felt when my mother proposed that Joel Miller adopt me. He had thought it was a great idea at the time, as my name would no longer be Goldfinger, and so my political activities would no longer threaten him professionally. As the conversation went well enough, I invited him to my home in Sudbury to meet my wife, Arianne, and our three children—his three grandchildren.

Clarence told me about his second family. I knew that he had two more children: David, whom I had met as a baby, and Susan, whom I had never known. Susan had dropped out of school at age sixteen and was living with a boyfriend in San Francisco. Clarence asked me to meet her the next time I was there, as he thought I might understand her better than he did. I did meet with her, but I could not really help her. I think she had suffered with our father in some of the same ways I had.

Over the years, Clarence and I continued to talk and meet. Eventually, both Beverly and I were sometimes included in his family events. After my mother died, Beverly and I wanted to find a way to undo our adoption, but that proved impossible: for Clarence to become our legal father, regardless of biological reality, he would have to adopt us. The law required that our current legal father, Joel Miller, be notified so that he could attend the court hearing. Clearly, these rules were designed with minor children in mind. Joel was not well, and we wanted to do nothing to disturb or hurt him. After Joel died in 2009, Beverly and I arranged for Clarence to legally adopt us. We went to court to explain all of this to the judge. When Beverly, Clarence, and I arrived at the courthouse, the waiting area was filled with young couples holding babies. Unable to help myself, I shouted out, "Daddy, daddy! Pick me up." We met with the judge for a few minutes, explained the situation, and left his chambers father and children. Since I was known professionally as Miller and had children and grandchildren with that name, I kept it.

My father's wife, Eleanor, died in 2005. Clarence stayed active and fit, working out at the gym six days a week until he was ninety-four years old. His memory then began to fail him. He died in January 2020 at the age of ninety-six. He never once said he was proud of me. He did eventually say he loved me.

Those to whom I have told the story of my childhood often comment on how awful it was. Some have even described it as brutal. But that is not how I remember it. I remember my childhood with joy. I remember my mother's love and my father's humor. I remember all the curiosity, exploration, and discoveries I made along the way to becoming Avram Miller.

3. A FLEDGLING SCIENTIST

Technology today is so different than what I experienced growing up. Now, when I watch my grandchildren use their devices, I imagine what my childhood would be like if I were growing up today. Back then, my only device was a pencil and paper, my only knowledge of computers was that they were enormous and that people in white coats operated them.

Back in the 1950s, televisions, radios, and phonograph players were significant pieces of home furniture. I had my own radio, my first piece of personal electronics, a crystal set that I built from a kit. The crystal meant it did not need electrical power; it received just enough of a radio broadcast signal through the crystal to power an earphone. I could listen to radio programs when I was supposed to be asleep, and no one knew. Since I was very often sick and in bed, my radio was a welcome companion. I spent my days reading books and listening to the radio.

I first watched television at the Stanford Children's Convalescent Home, where I became a big fan of *Howdy Doody*, *Hopalong Cassidy*, *The Lone Ranger*, and later, *The Mickey Mouse Club*. When I returned home, my family did not have a set, but my grandmother Della did. We went to her house to watch television until we got our own. Then, when I was home ill, I sat in a La-Z-Boy chair in our living room wearing my pajamas and bathrobe and watched television. There were only three channels. I watched series like *Flash Gordon* and, of course, *Superman*.

I loved these programs. I began to dream about the future, especially since living in the present was not so pleasant.

At that time, televisions were massive, but with oval screens not much larger than an iPad. Images were displayed on the screen by a cathode-ray tube; all the electronics inside used vacuum tubes. It would be some time before the common use of transistors would dramatically reduce the size of televisions. The development of transistors in the postwar years would lead directly to the creation of Silicon Valley and companies like Intel decades later. Unlike transistors, vacuum tubes were not very reliable. If a television set stopped working, which was common, service people would come to your home to fix it, which usually meant finding out which vacuum tube was no longer working. I discovered that you could go to a hardware store and use their vacuum tube tester to determine whether or not a vacuum tube was working and then buy a replacement. Most of the time, the tubes would fail because air would leak inside the tube, letting the filament burn out. That was easy to identify. So, by the age of ten, when other boys were playing baseball, I had become a "TV repairman." My career in technology had begun.

MY CAREER STARTED IN A MENTAL INSTITUTION

All that experimenting with electronics as a child paid off. At the end of 1966, my friend Mel introduced me to Professor Joe Kamiya, who had a small laboratory on the third floor of the Langley Porter Neuropsychiatric Hospital, part of the medical school of the University of California at San Francisco. Joe would become one of my most influential mentors.

Joe was studying brain waves, electrical signals produced by the brain and recorded by an electroencephalogram (EEG). I'm not sure I even knew that brains produced electrical signals. Joe was interested in biofeedback, and he needed help to develop the equipment he required for his experiments. That's where I came in. I guess he had not yet started looking for someone in earnest. Mel had just completed a residency with Joe and knew he needed someone to help him build the needed equipment and introduced me to Joe.

My job would be to design the equipment required to make the experiments possible. The equipment acquired the brain wave signals, identifying and determining the dominant frequency from a subject. It then provided the subject with an audio signal that would help them know when they were generating alpha waves. Although I had little in the way of qualifications, I told Joe that I would try to build the equipment he needed without pay but that I expected him to hire me if I was successful. I was, and a few days later, he brought me on as a research assistant. I was able to quit my night job at the pizza parlor. Sometimes I think that had Joe not hired me, I might own a chain of pizza parlors today.

Biofeedback is very underappreciated, because we are often unaware of many of our bodily functions. If you think about your breath, you become aware of your breathing, and then you can consciously control it. If you notice your heart is beating fast, and then rest, you can slow it down. This is easier to do if you can monitor your heart rate directly. It is harder to control something like your blood pressure, especially without objective information, and harder still to control brain activity, of which we have very little awareness. Meditation practice is meant to increase such awareness. The electrical signals the brain produces relate to its state. For instance, during sleep, the brain produces slow oscillations, called theta waves, that are also sometimes produced during meditation. In a conscious state, the brain generates mostly beta waves, and it generates alpha waves in a relaxed state.

Joe postulated correctly that if we could identify signals that correlated with different brain wave states, we might learn to control those states. He wanted to study people in meditative states, like Zen masters, and see if he could teach other people, subjects, to mirror those brain states. Then he would ask those subjects to report their experience. Joe was right. It worked, as it turned out.

Joe had already purchased several logic boards from Digital Equipment Corporation, a company I knew nothing about that made products for implementing digital logic, about which I also knew nothing. The boards were branded "Flip Chip Modules." The manuals Digital Equipment provided with the logic boards were very well written. I started to understand not only how the boards worked but also the general principles of digital logic, which is the foundation of computers. The more I grasped, the more excited I became. I had entered the world of digital technology, beginning a journey that has not ended to this day.

Digital logic comprises the fundamental building blocks of computing, logic elements can be constructed from simple electronic components: transistors, resistors, and capacitors. The logic boards were installed in a unit that also housed the plugboard, something like a telephone switchboard (which you may have seen in old movies). There were twelve logic boards, each with, on average, about four different components. I programmed the logic functions I wanted using small cables inserted into the various holes in the plugboard to make connections among the various logic elements.

We live in a world comprised of both analog and digital elements. Your weight and height are analog. An example of digital is a traffic light that is either green or red. Analog devices can be configured to be digital, having one or more defined states, and digital devices can simulate analog devices. Digital logic is a set of rules, a form of arithmetic, that can be implemented by digital devices that act as switches. The most basic is a device called a "gate." Modern microprocessors have billions and billions of such gates. But back then, I was working with gates just one at a time, in total counts less than a hundred. Sometimes, I would actually build logic elements out of transistors, resistors, and capacitors, the raw, analog elements needed to assemble a gate. Working at such a fundamental level gave me a profound appreciation of digital technology, which I still have.

For me, the discovery of digital logic was akin to a religious experience. I learned about the Boolean system of algebra, developed by George Boole in 1847, in which ones or zeros represent all values. I spent nights at my kitchen table studying, overwhelmed by the capabilities of such a system. Very complicated computations could be performed, and digital logic could be used to describe elements of the real world. I learned to do arithmetic in the binary number system Boolean algebra implies, just ones and zeros, also known as base two (because there are only two possible states). I learned octal, represented by a set of three ones or zeros and often termed base eight (with eight possible states for each unit of three binary digits). The decimal system that most humans use, base ten, is excellent for counting with your ten fingers and ten toes, but it's not so good for computers. Every night after work, I would sit at my kitchen table and study. It was as if my brain was on fire.

Joe knew and taught me a lot about electronics. He knew the basics of digital logic, but I soon exceeded his knowledge and started studying books on symbolic logic, a form of mathematics. Together, we also developed miniature analog amplifiers that could capture the extremely low-level signals coming from the brain. We put an array of small electrodes on a subject's scalp, glued in place. That was pretty messy. The brain produces little power in the form of electrical signals, so we were monitoring just a few microvolts, more than a million times less than the power of a nine-volt battery. It was not easy to capture these signals. Then I created the digital logic to process these signals using the logic modules from Digital Equipment and other elements I had created.

Importantly, Joe also taught me about the scientific method and statistics. He took a skinny misfit without a formal education and turned him into both a scientist and an engineer. I still knew nothing about computers, though that would soon change. But without Joe Kamiya, I would not be Avram Miller.

Though Joe was never good at publishing his findings, we did get some popular recognition for our work—work made possible by equipment I designed. It was an early highlight in my career. We were on the *CBS Evening News* with Walter Cronkite, and various magazines wrote on our work, including *Time.* Some of our notoriety came from Joe's study of Zen monks. This was the '60s, after all. We even did some work for NASA. Some astronauts reported out-of-body experiences, and we offered advice on how to handle this. I made up a name for our project: Astro-Not.

Our work attracted the attention of some notable people. For instance, the renowned scientist John Lilly came by. He was one of the scientists to study communication with dolphins and a proponent of sensory deprivation tanks. Even Timothy Leary of LSD fame stopped in at Joe's lab.

PDP-7

MY FIRST COMPUTER

One year into my time working with Joe Kamiya, Noch Calloway, the director of research at Langley Porter, purchased a computer for the research organization. It generated a great deal of excitement the day it showed up. I only knew of computers from the television, and I didn't know much about them or how they worked. The machine weighed a thousand pounds. I had never seen a piece of equipment that size.

The night the computer arrived, I was working late in the lab, and no one else was around in what had just become the computer room. I was curious about the machine, a PDP-7 from Digital Equipment Corporation, the same company that supplied my digital logic boards. I went behind the computer, which had doors on its back. I opened a few. Surprised, I found stacks and stacks of the same digital logic modules I had been using to build equipment for the brain wave studies! So naturally, I looked for the plugboard, which I could not find. So how were these logic boards programmed? Intensely puzzled and curious, I started to read all the manuals that came with the machine and examined the schematics. Around three in the morning, I suddenly and viscerally understood: software! I would learn the name later. That night, I taught myself to program in assembly language, just one level up from the series of ones and zeros that told the computer what to do. I had become a programmer. Now I was both a hardware and software "engineer." This is unusual, as most engineers only have only one of these skills. The concept of software or programs was developed by some amazing people like John von Neumann and Claude Shannon just twenty years earlier. The first high-level computer language, Fortran, was less than ten years old. There were no computer study programs at universities. Programming is one of the most powerful forces. Its source is imagination and thought. Like music, it is created out of nothing but the mind. It is no wonder that having already composed music, I would be drawn to computer programming.

My education on the basics of computers was now complete. I could create gates from transistors, resistors, and capacitors. I could assemble gates into logic elements like flip-flops (switches), and I could assemble these elements to make computations. This provided me with a very strong technical foundation that would play a continuing role in my career.

The Langley Porter research department had no technical person to support the computer. I asked Joe if I could reduce my hours with him to work part-time in the general lab as the computer support person. He agreed. I had fallen in love with a computer, and it would not be the last time.

In equivalent 2020 dollars, the PDP-7 cost more than $500,000. Only 150 were ever built. Ours was number 145. The computer had an eighteen-bit word length, an unusual design from the perspective of our computers today, which use multiples of 8 bits (16-, 32-, 64-, and 128-bit words). Though its size meant it certainly was not a personal computer, that PDP-7 was *my* computer. Interestingly, the first Unix operating system was written on a PDP-7. Its main memory held a maximum of 144,000 bytes (characters), on the order of a hundred thousand times less memory than a typical computer today. The machine had only sixteen instructions, one of which, "no-op," did nothing at all besides tell the computer to proceed to the next instruction on the next clock cycle. I still find myself sometimes describing a person as a no-op. Hardly anyone knows what I mean.

Attached to the computer was a paper tape reader and a magnetic tape unit. The keyboard was a Teletype ASR-33, which could both read and write paper tape. There was no disk drive. The PDP-7 also had no permanent read-only memory (ROM). After turning on the computer, one had to enter into the memory thirteen instructions via eighteen console switches, comprising a program called the bootstrap or RIM (read in memory) loader. That program allowed another small program to be read via the high-speed paper tape reader. This process, called "bootstrapping," took at least ten minutes. Now you were at a point where you could finally begin to use the computer.

Programs were written in the computer's specific assembly language, which translated written commands into the ones and zeros used by the computer. I still remember many of the PDP-7's sixteen instructions. I used a program that came with the computer, called the editor, to write programs. Then, I would assemble these programs using another program, called the assembler, that converted assembly language into the machine's instruction set. I would load the resulting program and hope it would run. If I made a mistake, there were few good options. Debuggers, which allow you to make small changes to a running program, came later.

I learned a great deal about computer programming over the following year. I implemented statistical processing of data from various experiments, as Joe Kamiya taught me about statistics. This greatly helped me in both my career and my investments.

The PDP-7 had an analog-to-digital converter that could take in signals from tape recordings and turn them into numbers that the computer could process. However, doing this conversion required the computer to run in real time, which complicated things. It meant that certain activities of the computer would have to take place at defined times. I discovered that the PDP-7 had a design flaw that made it difficult for it to function in real time, related to the way it processed interruptions. I figured out how to change the computer's internal code, something we now call microcode, to fix this problem, and I provided this information to Digital Equipment. They incorporated my design in their next-generation computer, the PDP-9, which I later used myself when working at the Thoraxcenter, at the medical school of Erasmus University in Rotterdam.

Pete Harris, the son of the executive director of Langley Porter and a few years younger than me, also became interested in programming. He got a job working alongside me. We had a great time and would often work all night. When we discovered that the lights on the computer console, when activated, interfered with our transistor radios, we realized that we could program the lights to flash in such a way that they would generate radio waves as they turned on and off at a very high speed. We used this capability to create and play music through our radios. The tape drive would vibrate when the tape moved or stopped. We learned to program it so that we could get it to "dance." There was also a CRT display, which we normally used to plot data. Pete figured out how to get it to write characters. There we were in 1967, making music, putting up lyrics, and having the magnetic tape unit dance. I loved that computer.

In 1968, I applied to take a civil service test so I could be qualified to become a systems analyst and be paid more. I was told that I could not take the test because I had not taken university-level calculus. However, because of my expertise in computing programming, I was later asked to join in the administration of the test I was not allowed to take.

MORE MONEY IN HEARTS THAN IN BRAINS

Late in 1966, Ronald Reagan was elected governor of California, which, unsurprisingly, I found very upsetting. But then his election affected me professionally in 1968. Joe had research funding from the Federal Department of Health for some of his projects. However, the money was administered through the California State Department of Mental Hygiene. Somehow, Reagan found a way to freeze this spending even though the funds were not formally California's to allocate.

A friend was doing some work at the Pacific Presbyterian Medical Center in respiratory monitoring. His former boss, Jerry Russell, had a PhD in biomedical engineering. Jerry was not a computer person, but he had recently moved to Rotterdam to join Dr. Paul Hugenholtz in creating what became one of the leading medical institutions in Europe, the Thoraxcenter, which was associated with Erasmus University and the Dijkzigt Hospital in Rotterdam.

I decided to write to Jerry and tell him that I was interested in learning more about his work at the Thoraxcenter. I suggested that perhaps there could be a role for me there. I was not that serious, because I was happy working for Joe and enjoyed my life in San Francisco even though I was frustrated with Reagan's actions. Before I knew it, I was receiving calls at home from Professor Paul Hugenholtz. Paul wanted to make computers an integral part of patient care. He said that the Thoraxcenter needed someone with my skills and vision to lead its effort. Evidently, he had already heard of me from people at Digital Equipment Corporation. I was just twenty-three years old, with barely a high school diploma, but was getting noticed.

I was attracted to the potential ability to work with state-of-the-art computers. The Thoraxcenter had already decided to acquire two PDP-9s, the successor to the PDP-7 I was using in San Francisco. Though I didn't want to leave Joe's lab, our resources were minimal. It seemed there was more money for hearts than brains. I decided to leave Joe's lab and join the Thoraxcenter, which meant moving to Rotterdam. I would live outside the United States for the next ten years.

By March 1969, I was bound for Rotterdam with a few stops along the way. First, I had some training sessions at the headquarters of Digital Equipment Corporation in Maynard, Massachusetts, in the Greater Boston area. Digital's headquarters were an old Civil War–era mill, which naturally Digital called the Mill. Mort Ruderman, who arranged my trip, ran Digital's medical product line and would become both my friend and my mentor. He gave me a tour of the headquarters. Little did I know that, just ten years later, I would work in that very same building.

The winter before I started my new job, Paul invited me to stay at his home in Massachusetts. It had been snowing heavily when I arrived. Growing up in San Francisco, I barely knew what snow was. I started across the yard as Paul opened the front door. Before I knew it, I was waist deep in snow and very embarrassed. And so, Paul and I met face to face for the first time.

I arrived in Rotterdam on March 24, 1969, after stops in Paris and London to visit friends and family. After checking into a hotel, I went to Paul Hugenholtz's office. The Thoraxcenter was still under construction, so the staff were located in temporary buildings. The arrangements for my travel and relocation had been made by Paul's assistant, Arianne van der Klooster, with whom I had corresponded. I knew nothing about her: how old she was, what she looked like, or whether she was married. As I discovered, Arianne was just twenty years old and beautiful in a delicate and refined way. Like many Dutch people, she spoke many languages. She welcomed me in perfect English.

Arianne showed me around the facilities. That night, she took me out to dinner and gave me a bit of a tour of Rotterdam. It was not a date, but at the end of that evening, I asked her to join me for dinner the next evening. That, I made clear, would be a date. Three weeks later, we were engaged, and seven weeks later, we married. Arianne's position working for Paul, my boss, was now clearly awkward, so we decided she would resign. Paul was not very happy about that, but it did not seem to harm my relationship with him. Arianne and I had two sons and a daughter together. We were married for twenty-five years and remain close even today.

Soon after marrying Arianne, I decided to learn to speak Dutch. It was no easy task. I had the Berlitz school design a six-day intensive course for me. I had five different teachers and studied ten hours each day. Arianne had agreed that we would only speak Dutch at home. After the course, I could get by well enough; within six months, I was fluent. Arianne and I continued to converse in Dutch for all twenty-five years of our married life. Today, twenty-five years after our marriage ended, I am still fluent in Dutch.

Me with the PDP-9; note that the name was Appel (Apple).

At the Thoraxcenter, I was appointed head of the computer department, which actually meant head of two computer departments: one at Erasmus University and one at the Dijkzigt Hospital, both located next to each other in Rotterdam. I came up with the name The Department of Clinical and Experimental Information Processing, which has lasted to this day more than fifty years later. When I arrived, two employees had already been hired to work on the computer. Mr. Bruin was much older than I was, and he was perhaps in his midforties. Cees Zeelenberg, a young engineering graduate, who would take over my position five years later, was the second person. He had a very successful career in medical computing and became a professor. All three of us sat in one large room. The computers were not yet operational. They stood in what would be the computer room, covered in plastic.

As it turned out, Mr. Bruin was not suitable for the kind of programming I envisioned for the group. I decided he would have to go, but I had never fired anyone before. It was not such an easy thing to do under Dutch law, anyway. I discussed it with Paul, and he agreed. I had to tell Mr. Bruin, a man almost twice my age and with a family to support, that he was terminated. It was very hard, and I cried afterward. Learning to let people go is an essential skill in business. We all make mistakes in hiring, which are necessary to correct. One should always let people go with consideration and respect, however.

In addition to developing software, we integrated different kinds of hardware and equipment, some of which we designed. I discovered Wim Englse, one of the most talented engineers I have ever met. Though he only had a vocational degree, I recognized his genius right away. I had to hire many software, hardware, and operations people. We had to go to England to get most of the programmers, as the Dutch computer industry was not well developed. I also brought Pete Harris over from the United States; he was a big help getting things going and stayed for about a year. I built a good team on the foundation of Cees, Wim, and myself. We had an incredible five years working together. Paul let me bring an administrative assistant from the United States, Mary Anne. A former marine, she was pretty tough. I once had to bail her out of jail when she got arrested after a bar fight. Soon after that, she went back to the United States. I hired a wonderful Dutch woman, Mop van Nierop, to replace her. Eventually, my group had about thirty-five people. Still in my midtwenties, and still with no postsecondary education, I now held an academic appointment at the university equivalent to an assistant professor. Later, I would become an associate professor.

Our most significant project was to create a monitoring system for the intensive care and coronary care unit, for, I think, a dozen beds. One of the challenges was to develop a system that was reliable enough to run twenty-four hours a day, seven days a week. At that time, it was unusual to run a computer continuously. Computers, in general, were incredibly primitive by today's standards. We found a video disk system that we could program bit by bit to create characters. This was one of the first bitmap graphics systems ever implemented, with output displayed on television monitors turned ninety degrees so that they looked like a piece of writing paper. Initially, we used a rotary dial from telephones as input to control the system. Nurses could enter numbers corresponding to various functions by dialing them in. Later, we built a keyboard with just ten digits and a few control keys.

We had two identical PDP-9s from Digital Equipment: one for monitoring, and one for programming and to back up the monitoring computer. The process of programming was quite primitive back then. We mostly wrote in assembly language, and sometimes Fortran. The programming system allowed for only a single user at a time. We had at least ten programmers, so we had to allocate computer time among them. I loved programming, but as my group grew, I had less and less time for it. Eventually, I lost my programming privileges altogether. That was one of the saddest things that happened to me. Programming is one of the few opportunities we have to create something out of nothing.

We developed a program to look for arrhythmia (irregular heartbeats) in EKGs, which required a lot of computing, which was beyond the capacity of the PDP-9. We augmented that machine with a PDP-15, the successor to the PDP-9, which had no peripherals. We created a high-speed network between the PDP-9 and the PDP-15 so we could send the EKG information to the PDP-15 and get back analysis. I began to think about computer networks, and I even gave a talk on the topic to the Dutch Computer Association.

Before using the systems with patients, we tested them with pigs, which have a very similar cardiovascular system to humans. We got a bunch of pigs, and the surgeons operated on them to induce various problems. I was sometimes present in the operating room during such procedures, scrubbed in and wearing the same white clothing as the surgeons. My first time in, one of the surgical team said to me, "If you feel faint, fall backward." In other words, don't fall into the pig.

THE KILLER APP: DIGITAL TIME

The first day the system became operational with patients, I was so excited. I could see that the nurses were using the system, which made me very happy. I finally walked over to the nurse's station and asked if they liked the system. One of the nurses said it was terrific. I asked, "What do you like the best?" She replied, "Seeing the time on the display." The nurses eventually came to appreciate the system's more advanced capabilities to help them care for their patients.

A young researcher at the Thoraxcenter, Klaas Bom, not yet thirty years old, was a pioneer in the use of ultrasound to measure heart function. He was a principal developer of what is now called echo-cardiography. My group developed a computer system to help analyze the data. Years later, in 2003, my second wife, Deborah, and I had several scans done at Cedars-Sinai Medical Center in Los Angeles. I had donated substantial money to Cedars in 2000 to help set up the Prostate Research Center under the leadership of Stuart "Skip" Holden and David Agus, so we got the VIP treatment. The head of the imaging group reviewed my coronary scans with me. I mentioned my role at the Thoraxcenter, and he was amazed, telling me how important that early work had been and how outstanding the Thoraxcenter had become. He then asked if I had time to join him in his research lab. He treated me like a rock star. I had frankly not thought about that early imaging work in many years. I was proud, of course, that I had been involved and had made a contribution.

Paul asked that my group develop a real-time computer system for the Thoraxcenter's catheterization lab (cath lab), which was led by Dr. Geert Meester. Cath labs are facilities in which investigations into a patient's heart function can be performed. The most common procedure is a coronary angiogram. Radiocontrast agent is injected into a patient's vein so that it is possible to visualize the flow of blood feeding the heart itself, for instance.

We decided to use the PDP-11, Digital Equipment Corporation's first 16-bit computer. That meant the PDP-11 had a different architecture and instruction set from the 18-bit computers we were already using for the intensive care unit. We correctly identified that those 18-bit computers were not likely to survive in the future. Of course, at that time, I did not realize that just eight years later, I would be responsible for leading engineering on all low-end PDP-11s at Digital.

We not only developed a computer system but also did some groundbreaking imaging work. We acquired a video disk system, intended for instant replay during sporting events, that could play back thirty seconds of video, including in slow motion. That allowed the doctors doing the procedure to watch the X-ray video sequence repeatedly. Before this, they had just one shot to see how the dye they injected was moving within the arteries of the heart. Then, we had the idea that we could subtract the background, since it was not changing. That made the flow of blood much more visible.

We developed a simulation, or model, of the electrical properties of the heart. With the limited computational power available then, it took more than a day for the model "heart" to beat just once.

About three years after I began my work at the Thoraxcenter, I was offered the possibility of getting a PhD from the Delft University of Technology, the primary engineering school in the Netherlands. The Thoraxcenter had close ties with some of the faculty there, and I was pretty well-known by that time in the biotech community. It was possible to submit a PhD dissertation compiled, more or less, from academic papers the PhD candidate had written, even if not as first author. And I had shared authorship on many already-published articles, some in prominent medical journals. After much consideration, I declined this opportunity. First, I was uncomfortable using the papers I had written with others. Second, I was afraid I might lose my edge if I became "legitimate." I liked the notion that I was so successful in the academic world without ever having attended university.

Not long after I joined the Thoraxcenter, I realized I had a lot to learn from Paul Hugenholtz. I wanted him to be not only my boss but also my mentor. He was a very busy man, so it was hard to get quality time with him. Then I realized that once each week, he drove an hour to Amsterdam, where he gave a lecture, and then drove back. If you were alone in a car back then, there was not much to do besides listen to the radio. So, one day, I stood in front of his car before he left the Thoraxcenter to drive to Amsterdam. I asked him if he wanted company. "Why not?" Paul replied. So most weeks, I got two hours of quality time with Paul, who was a great physician and scientist, as well as a leader with great charm. I wanted to imitate him.

I was very active in the Digital Equipment Computer Users' Society (DECUS), a user group for those using Digital's computers. In 1972, I organized a meeting at the Thoraxcenter for members of the Dutch user group and Gordon Bell, the head of engineering at Digital Equipment. After the meeting, I drove Gordon to Schiphol Airport, an hour from Rotterdam, so we had plenty of time to talk. Gordon was in his late thirties at the time. Less than ten years later, I would work under him as group manager for the Professional series at Digital Equipment.

4. ALIYAH

I had originally planned to stay at the Thoraxcenter for only two years. After I married Arianne, circumstances changed quickly as we had our two sons. After five years, though, I felt it was time for me to do something else. We were seriously considering moving to Israel when Dr. Paul Hugenholtz was tapped to lead the development of a significant outpatient facility at Dijkzigt Hospital. He had done such a fantastic job of creating the Thoraxcenter from scratch that the hospital's board thought he could do something like that again. They wanted him to revolutionize outpatient care, especially by integrating computer technology.

Paul's new project had a considerable budget, and he was convinced that I was the right person to lead the technology development efforts. I asked, "Does this mean that I have finished my job at the Thoraxcenter?" He agreed, "Yes, you have done everything you said you would do, and now you can leave it to someone else." I said, "If I am done, I am going to quit and move to Israel with my family." I even surprised myself. I had no idea I could say something like that. Paul was in shock, I think, but he didn't try to change my mind. That night, I told Arianne, and she agreed we would immigrate to Israel. I had already been offered a position at Mennen-Greatbatch there, an offer I accepted a few days later. Six months would still elapse before Arianne, our two boys, and I made the journey to Israel and became

Israeli citizens. In the interim, Israel fought the Yom Kippur War in October 1973. The war only strengthened my resolve to move to Israel.

ZIONIST

Looking back, my desire to move to Israel was deeply rooted. I was—and still am—a Zionist, and I believed that Jews needed to have their own nation and would not survive in the long-term otherwise. I always thought of myself first as a Jew who lived in America rather than an American who was Jewish.

Two of my grandparents were born in the United Kingdom, but their parents were originally from modern-day Poland. Another was born in Poland but left before the age of one. My mother's mother, Della, was born in California, soon after the 1906 San Francisco earthquake. Her mother, Bessie, my great-grandmother, lived to ninety-seven, and she was around a lot as I grew up. She sometimes watched after and cooked for Bev and me, and she told me many stories about her childhood in Priluki, Ukraine. Bessie came to San Francisco in 1897 after a very long journey, where she married Abraham Levy, sixteen years her senior. He died in 1928, and I was named after him. It is a Jewish tradition to name children after relatives who have passed, but often that just means adopting the same first letter of the name. Bessie lived long enough to hold my eldest son, Adin, in her lap. I have photographs of all five generations together. She was a character. As she grew very old, she would sit in front of the TV and scream in Yiddish at the people on the screen. When I asked her to speak English with me, she would often answer in Yiddish telling me that she was speaking English.

A Jewish youth organization had also helped me develop as a leader during a very formative time. At thirteen, I was a skinny, withdrawn kid wearing glasses. My mother, concerned about my lack of social contact, suggested I join a Jewish youth organization for boys, AZA (Aleph Zadik Aleph). Reluctantly, I agreed. I was very reclusive and uncomfortable being around other children, but that would soon change.

B'nai B'rith, one of the most important Jewish organizations in the United States, established AZA in 1924 (B'nai B'rith itself dates back to 1843). In 1944, they formed a girls' organization, BBG (B'nai B'rith Girls). These organizations set up chapters, like fraternities and sororities. Larger cities could have several chapters; as I recall, San Francisco had four or five for each gender. The fraternal organization's stated guiding principles were patriotism, Judaism, filial love, charity, conduct, purity, and fraternity. Members were aged from thirteen to about eighteen, and each chapter had a set of officers: president, vice president, treasurer, and the like.

My chapter, King Solomon, met about a thirty-minute drive from my home in the upstairs of a building that was owned by a synagogue. The girls met downstairs. We would have social gatherings together or stop on the way home for a snack at Zim's, a hamburger place, since many of the members were old enough to drive. My mother's plan worked; I felt accepted as a friend. I met my first girlfriends through BBG, which my sister Bev joined when she turned thirteen.

AZA was crucial to the course of my life, because there I became a leader. Though I did not know it at the time, AZA would prepare me for corporate life. I was given some responsibility right from the beginning, appointed vice president midyear because one of the officers moved away with his family. Then, in the next formal election, I became president of the chapter, even though, at fourteen, I was several years younger than the other members. I served on the regional board, and I learned leadership skills that I would use throughout my career. All of my children would later be deeply involved in the B'nai B'rith youth organization and obtained even more significant positions than I had ever achieved at the international level.

By that time, my mother and stepfather owned three donut shops in San Francisco. A few of my fellow chapter members worked at the stores after school, and my mother would often bring donuts to chapter meetings. Then, when I was fourteen years old, their donut shop on the corner of Irving and Twentieth Avenue caught fire, extensively damaging the apartment building above it. Without adequate insurance, my mother and stepfather were sued, which drove them into bankruptcy. They lost everything but our home. My stepfather, Joel, got a job driving a delivery truck for Wonder Bread. My mother worked as a waitress. I quit AZA, too embarrassed to see my friends again, knowing that they knew about the bankruptcy.

Much later, when I was living with Arianne in Holland, I began to realize how much my Jewish identity really meant to me. Our two sons, Adin and Asher, were born in 1970 and 1972. The Nazis had killed more than 90 percent of the Jewish population in the Netherlands; only about eight thousand remained. I felt that the country was a Jewish graveyard. In Rotterdam, there were almost no Jewish families. We did not want to raise our children in this context. More than that, though, I wanted to be Israeli. Though Arianne had not yet converted to Judaism, she nevertheless felt she was a Jew. She was converted later, by the chief rabbi of Israel.

MENNEN-GREATBATCH

In 1973, I got an opportunity to fulfill my longtime dream of immigrating to Israel. One of the suppliers to the Thoraxcenter was Mennen-Greatbatch. Bill Greatbatch, one of the company's two founders, had invented the pacemaker. The company had a subsidiary in Rehovot, Israel, called MG Electronics and run by Moshe Barone. We, at the Thoraxcenter, had discussed with Mennen-Greatbatch the possibility of licensing to them the catheterization system that my group had developed. Moshe contacted me and asked if I would like to visit Israel and present the system at their sales meeting to be held in May. He told me that I should bring Arianne and that we would have a few additional days to explore Israel.

We landed on the evening of Lag BaOmer, a unique religious holiday. It is tradition to light bonfires that evening. As we flew into Ben Gurion Airport, we saw hundreds of bonfires burning on the beach. It was incredibly beautiful. We went with all the other attendees of the sales meeting to a large Moroccan tent in the countryside and ate a fantastic meal. The next day, after my meetings, we were invited to Moshe's home with a number of the others. Moshe was very successful, and his home reflected it.

The next day, I once again visited the company's office in the Kiryat Weizmann Industrial Park, in Rehovot, about thirty minutes from Tel Aviv. I discussed my interest in immigrating to Israel with Moshe. I told him I planned to explore getting an academic position at one of the universities. Moshe said that Israel needed no more professors. Instead, Israel required businesses, especially export-oriented businesses like MG Electronics, which did about 80 percent of its business outside of Israel. I had worked very hard to achieve academic success despite having no formal degree. Having consulted for Digital Equipment Corporation, Hewlett-Packard, and other companies, I did have some sense of business at that time, but much would be new for me.

Moshe suggested that I come and start a medical computer systems division not only for MG Electronics but also for its American parent company, Mennen-Greatbatch. The products would use the software that my group had built at the Thoraxcenter; we would work out commercial arrangements to compensate the Thoraxcenter through a licensing agreement. Israel was not a comfortable place at that time, nothing like the Israel of today. He offered to pay me a US-equivalent salary, about five times an Israeli wage for the same position at that time.

Through Paul Hugenholtz, I had arranged in advance to meet Professor Henry Neufeld at the medical school of Tel Aviv University and the hospital of Tel HaShomer. Professor Neufeld was the most renowned cardiologist in Israel, with a world-class reputation. We spoke about medical research, and I told him about my interest in immigrating to Israel. He offered me an opportunity to join his staff if I decided to come. But I had made the choice to pursue a career in business, both because of my financial needs and because I believed I could do more for Israel as a businessman than as a scientist. After I immigrated to Israel, I did eventually accept a position as an adjunct associate professor in the Department of Cardiology. So, at twenty-nine, I became an associate professor without ever having attended university.

I returned to Israel from Rotterdam to negotiate my deal with MG Electronics in early November 1973, just a few weeks after the end of the Yom Kippur War. Cars were still not allowed to use headlights.

I stayed in Moshe's home, and we met with members of the Chief Scientist Office of the Department of Commerce to line up government support for our efforts. Roughly, that totaled $3 million in subsidies ($10 million in 2020 dollars).

I have made many bad decisions in my life, and the worst was accepting Moshe's offer to join MG Electronics. While Moshe Barone was very persuasive, the decision was mine, and I have to admit money played a significant role. Sure, I wanted to immigrate to Israel with my family, but I didn't want to deal with what that meant. Israel had a system for integrating immigrants. Normally, Arianne and I would have attended an ulpan to learn to speak Hebrew. I would have gotten assistance finding work. The government would have given us loans and money until I found work. But I elected to work outside that system, and so I went to work right away.

Not leaving the medical field was another big mistake. At twenty-nine, I was already well-known. Having accomplished so much without even attending a university, I did not want to take the chance of starting over again in some other field.

When I visited in November 1973 to negotiate my position, I also investigated housing. Most Israelis lived then, as now, in apartments. Arianne and I owned a three-bedroom home in Rotterdam, which we'd bought when I was just twenty-six. I was so proud that at a young age, I could provide for my family like that. I could not imagine living in a typical Israeli apartment. Moshe lived in Savyon, one of the wealthiest communities in the country at that time, if not the wealthiest. The houses were large, with big yards. There was a small shopping center and a country club. I found a development of new homes on the edge of Savyon that would take another year to complete. While they were not nearly as expensive as most homes in Savyon, I still lacked the funds to purchase one. Between my savings and the money we expected from the sale of our home in Rotterdam, plus a loan from the Israeli Ministry of Absorption, I could pay for half of the place. Moshe proposed that the company purchase the other half. We would be joint owners, and I would not have to pay the company any rent.

I called Arianne and told her about the house. She agreed that we could buy it without seeing it. (In those days, I could not email her a photo.) What I neglected to consider was that this deal made it impossible for me to leave MG; I would have no way to come up with the other 50 percent of the home's cost.

Arianne, our two sons, and I moved to Israel in June 1974, moving into a home to rent nearby. We lived there for a year while the house we bought was completed. A few years later, our daughter, Dafna, was born.

Building near our computer room

MG's building at the Rehovot Science Park was too small for my new group. We found an abandoned moshav (collective) about half a mile away and rented it. I ordered a PDP-11 for our computer lab, a large, one-room building. A small group of Bedouins lived in front of it. Every day, I found myself stepping over several women and children in their traditional garb to enter this modern computer lab. A year later, once the main company building had expanded, we moved there.

I found a programmer who had just graduated from the nearby Weizmann Institute, which was and is one of the most excellent research facilities in the world. Established in 1934, the institute only offers graduate and postgraduate degrees in science. More than six Nobel laureates have taught there.

He introduced me to several other students, whom I also hired. I hired people to make both hardware and software. I even hired a Palestinian who had trained in the United States. That became a problem for the Israeli authorities, because we were in the export business and shipped a lot of our goods by air. We had to ensure he got nowhere near our shipping area. I also had many problems finding him housing in Rehovot. Eventually, he left, frustrated, and moved to Jerusalem. It was a failed experiment. We also saw a flood of talented engineers leaving the Soviet Union and coming to Israel. One told me that he had programmed a Russian-made Minsk computer and that he could only test his programs twice a year. I thought this would make him very cautious and unlikely to make mistakes, so I hired him.

We licensed the cath lab system that my former group had developed at the Thoraxcenter. Mennen Medical already had a coronary care unit product, called Argus. We took over its development. A few years later, we designed a single-user computer, called Solo, using a single-board computer that Digital Equipment Corporation had recently launched.

A TRAVELING SALESMAN

Before my job at MG, I had never really dealt with business matters. My focus had been on technology development. Mort Ruderman, my friend who had run the medical product line at Digital Equipment and later founded Meditech, had encouraged me to immigrate to Israel. So, in return, I asked him to come to Israel for a few weeks and help me understand how to run a business. He agreed, for which I am still very grateful. He taught me about profit and loss, balance sheets, budgeting, and sales processes. I felt like I got an MBA over a few weeks' time.

I had much to learn about pricing and sales terms. I hired a marketing person, but he was not much help. Moshe was also very involved in all of the business decisions. Somehow, we soon started selling out of the cath lab and ICU systems. As it turned out, I was a great salesperson. Because MG Electronics was a subsidiary of an American company, we conducted all business in English. The domestic market was small, so about 90 percent of our customers were outside of Israel. I was also responsible for the parent company's product development of computer systems. Between that and sales, I traveled a great deal. I was probably on the road more than half of the time. We installed a fax machine in my home so I could communicate with our distributors. We would type communications out by typewriter and then send them by fax. No email in those days.

While traveling so much was very difficult, especially for my family life, it was also fascinating.

I went to South Africa, which I found to be interesting, beautiful, and deeply disturbing. Apartheid would continue there for another twenty-five years. Here I was, a former civil rights demonstrator, traveling and doing business in apartheid South Africa. I justified my involvement by making sure we donated equipment to Black hospitals. Really, though, I wanted to see what was going on for myself. I could not help being curious. I fell in love with South Africa, especially Cape Town. Our distributor there was about my age and Jewish. We traveled together throughout the country. White people were very privileged, especially those I typically met socially, and they often employed several Black servants. Staying at a friend's house once, I learned that the woman who brought my morning tea left her village and traveled more than three hours to reach my friend's home by the required six a.m.

White South Africans fell into two major groups based on their colonial history: those who had originated in England, and those who had originated in Holland, called Afrikaners. I was very popular because I was Israeli, and Israel, I have to regrettably say, supported the apartheid regime by selling them arms. The two groups of white people spoke two different languages, English and Afrikaans. Since the latter is a derivative of Dutch, which I spoke fluently, I learned to understand and speak basic Afrikaans. That meant Afrikaners loved me, an Israeli who spoke their language. The Afrikaners were hard-core racists. When they spoke in English, they appeared moderate; in Afrikaans, their anti-Black feelings seemed clearer and firmer. That was hard to take. Decades later, I went back to South Africa several times. Apartheid had ended and Nelson Mandela was leading the country. The country has dealt with its segregated past better than most.

I visited Iran, another very interesting country. As with South Africa, Israel had a strong relationship with Iran, even providing mercenaries to protect the shah. I could fly to Tehran directly from Tel Aviv. I stayed at the Hilton Hotel, where many Israelis rented rooms to meet with business associates from hostile Arab countries. Israeli and Arab businessmen would transact sales at that time by shipping via neutral places, like France. At MG, too, we put "Made in France" on our equipment when we sold to customers in Arab countries. Tehran is located at high elevation near some unbelievably beautiful mountains and close to the Caspian Sea. Through our distributor and some of our customers, I was invited to social gatherings of wealthy Iranians. I learned to drink vodka straight and very cold while eating caviar. The word "opulent" comes to mind. In early 1979, the shah was overthrown in a bloody revolution. We learned from one of his employees that our distributor had been lined up and shot dead.

The Israeli tax authorities were very pragmatic. They knew that bribery played a role in doing business in various parts of the world. Shocked, I learned that a published tax schedule allowed companies to deduct a certain percentage of their sales price for bribes. I think the tax code used a nicer word than that, but I can't recall what it was. (Until the late 1990s, Germany, among other countries, also allowed this.)

I sold all over the world. We even supplied a twelve-bed unit to the Kremlin for top officials of the Soviet Union. We had enough Russian speakers in the company to translate the documentation into Russian.

We were doing very well for a time. But the business's scale was fundamentally limited, because I was the only one who could sell. I tried to get others to help with sales without much success, and my absences hurt in product development. After my first year, I had to give up my position as adjunct associate professor. I did not have enough time to do research, manage product development, and travel around the world selling our products.

A FAILED ZIONIST

All of the travel and the pressure took its toll. My family life was terrible, and my marriage was in crisis. I was never home. Arianne and I had agreed that we would continue to speak Dutch with each other and our children. Our boys, Adin and Asher, did not speak English. Asher could barely speak Dutch. But Arianne, very talented with languages, was soon fluent in Hebrew. She began to speak Hebrew with the boys, which was not consistent with our agreement. This resulted in their only speaking Dutch with me. Adin was only four and Asher two when we moved to Israel. Since I was rarely around, their Dutch did not develop. Our daughter, Dafna, who was born in Israel in 1976, never learned to speak Dutch at all. My Hebrew was almost nonexistent. I had a real language problem: I could not communicate with my own children.

Under a great deal of pressure at work and at home, I started seeing a therapist, who put me on tranquilizers to deal with my anxiety. That was not helpful. Instead of being anxious, I was now depressed. Just thirty-three years of age, I thought my life was over. It was far from over, but I can still feel the despair that would cover me at that time. This was the lowest emotion point in my life.

Arianne was miserable in Israel. She felt isolated. Israel was not the vibrant and exciting country it is today. One day, in the summer of 1978, she returned to Israel after visiting her family in Holland with our children, and when I greeted her, she broke down in tears. She said she could not remain in Israel. It became too much for us, and we decided to leave. I was crushed by that decision. My lifelong dream had been to live in Israel and to bring up my children there.

For the first time in my life, but not the last time, I felt that I had failed. I had just come to Israel as a Zionist and wanted to contribute to the country. But in retrospect, my contribution would have been much more significant had I started a computer company, for instance. Israel was not then the high-tech center it is now. I could have played a major role in laying the foundation for the Israeli technology industry. I eventually returned to live in Israel in 2017. Both Israel and I had undergone a great deal of change in the thirty-eight years since I had left. Israel was now the "start-up nation," the best location for start-ups outside of the United States. Intel is the largest private employer in the country, so my having been an officer there is respected. I have to wonder what would have happened to my life and that of Arianne and our children if we had stayed in 1979.

PART II

THE HEIGHTS (1979-1999)

*From Digital Equipment and
Franklin Computer to Intel.*

5. THE RISE OF PCS

By 1979, I had spent ten years specializing in the use of computers in real-time applications in cardiology (coronary care unit monitoring, catheterization lab). Thinking about my future and about leaving Israel, I felt that I was overly specialized and needed to focus on either medical science or computer design. If I chose medical science, we would return to Holland. I was well-known there, and it was Arianne's home and the place where my sons were born. I spoke fluent Dutch, so our family's language issues would disappear as soon as the kids learned to speak that language. If I wanted to work in the computer industry, however, we needed to move to the United States. Forced to make a choice, I realized I could never give up computers. While I enjoyed medical science, I was really at heart an engineer and wanted to be deeply involved with the design of computers.

At that point in my career, I made a critical decision, and in many regards, a very risky one, to leave the medical field and join the computer industry. What I did not know, of course, was the computer industry itself was about to go through a major transition in just a few years, driven by the rise of the personal computer, a transition that would have a profound impact on the world and would dramatically affect my future.

Although I contributed in only a limited way to the early development of the personal computer (PC) industry, I do have a unique perspective resulting from my decades in the industry. Thus, before

continuing on with my story, I want to pause here and recap the story
of the IBM PC.

IBM ACCIDENTALLY TURNS THE
COMPUTER INDUSTRY ON ITS SIDE

In 1964, when IBM introduced the System/360 line of mainframe
computers, it was primarily a tabulation and typewriter company.
This "bet-the-company decision" sure paid off. In just a few years, IBM
became the leading computer company in the world, capturing more
than 70 percent of computer-industry revenues. Strangely, a company
that made its fortune by developing a line of compatible computers at
different price points, all running the same software, would end up
damaged by that very same phenomenon of software compatibility.
This pattern of not recognizing discontinuities—even by companies
that had previously created them—would be repeated over and over.

— listen to the market

The story of the IBM PC is not primarily about hardware. It is a
story about software, as well as unintended consequences, hubris, and
greed. Nobody at IBM imagined that their own actions would transform the computer industry in such a way as to significantly diminish
their leadership role and even put the company's existence in peril.
Likewise, nobody at IBM foresaw that many small companies would
form, each taking a bite out of IBM, let alone that among these thousands of piranhas would be the likes of Bill Gates and Andy Grove.

Sometime in 1980, the top management of IBM, watching Apple
Computer's growth, concluded that IBM should enter the nascent PC
market. Apple had transitioned from serving a hobby market into building a productivity tool, driven by business applications like VisiCalc,
the first spreadsheet application. If the PC were to create a new market, IBM management wanted to dominate it. Atari, then the leader in

computer gaming and a subsidiary of Warner Communications, had contacted John Opel, IBM's president, with an offer to develop a PC for IBM to serve the business market. Frank Cary, IBM's CEO, asked Bill Lowe, the general manager of its Entry-Level Systems Division in Boca Raton, to evaluate this possibility. A few months later, Lowe presented his review to the IBM management committee. He recommended that IBM acquire Atari from Warner and use a repackaged version of the Atari 800 computer as a starting point for IBM's entry to the PC market. He even demonstrated to the committee a modified Atari 800, which had been brought to market a year earlier and was doing well.

My close friend Steve Mayer, one of the founders of Atari, had been the lead developer of the 800. This computer was designed for entertainment and games, so why Lowe thought it could serve IBM's needs is a mystery to me. Perhaps he thought IBM could never develop a competitive PC in time to enter the market and so at least wanted to take the available opportunity. Or perhaps he cleverly manipulated the committee to give him the authority to move forward outside of the normal IBM processes.

The IBM executives wisely rejected Lowe's Atari proposal. However, they understood the challenges of creating new products within the IBM culture. Developing a new computer at IBM could take five years or more, as IBM's cultural focus on quality impacted both design and manufacturing. Indeed, Lowe's team had already spent almost four years developing the IBM Datamaster, a precursor to the PC.

IBM had numerous committees and procedures. Anyone could say no, and no one person could say yes. To counter this, Cary had already begun to encourage the creation of small autonomous units—"independent business units (IBUs)"—to counter IBM's sluggishness. He asked Lowe to form an IBU to develop a PC. Lowe returned to the Management Committee on August 8, 1980, with a plan to set up a small team to develop the IBM PC. Don Estridge, who had been the assistant general manager of Entry-Level Systems, became the head of the newly formed PC team, which also included marketing and sales. Lowe committed to bringing the IBM PC to market in an astounding one year. Therefore, he explained that his team would have to go outside of IBM for most of the critical components, including the microprocessor and the operating system. The Management Committee approved this plan.

One year later, on August 12, 1981, the IBM model 5150—the IBM PC—was launched during a splashy press conference at the Waldorf Astoria in New York City. Its price tag was $1,565 (about $4,400 today). No one at IBM or anywhere else understood that the computer industry had just been turned on its side. IBM had launched a near-fatal blow to its own business, creating a "virus" that would almost kill IBM and that would prove a mortal blow to most of the other existing computer companies.

In order to meet the one-year product shipping deadline, the IBM PC team used existing hardware components from other companies. Importantly, this included the microprocessor: the Intel 8088. IBM also licensed the computer's operating system from Microsoft. The decision to go outside IBM for most of the PC's hardware and software was probably the most consequential decision taken to date concerning the development of the computer industry. But the decision's impact was not understood by anyone at the time, certainly not by IBM, and certainly not by Apple, at the time the leading PC company. On the day of the IBM announcement, Apple ran a full-page ad in the *Wall Street Journal* headlined, "Welcome, IBM. Seriously," with the touch of arrogance only Steve Jobs could supply. At this time, Apple had about 20 percent of the PC market. Their share would drop below 4 percent just a few years later. They mistakenly thought they would be competing only with IBM.

PROJECT CHESS

The PC unit's code name was Project Chess, and the first computer was called Acorn.

Strangely, Bill Lowe was promoted right in the middle of the PC's development, and Don Estridge, his number two, was now placed entirely in charge of the effort.

— small teams can outperform big ones

The team was just thirteen individuals, and that included marketing, licensing, and software people. But the power of small teams should not be underestimated. Yes, they used off-the-shelf components and subsystems, but there were many, and they required integration. They also had to develop an overall architecture, create a third-party software business, and set up distribution, among other things. Robert X. Cringely, in his excellent book *Accidental Empires*, explained their speed as a head start; they reused much of the work already done in developing the IBM Datamaster, a word- and data-processing system aimed at small businesses. The Datamaster, released a month before the IBM PC, used the Intel 8085, which had a compatible bus architecture with the 8088. Still, I scratch my head. None of the people I met who worked on the team really impressed me, with the exception of Don Estridge.

BRING IN THE CLONES

Many think that Microsoft and Intel created the PC revolution, but Compaq actually set the course. When the IBM PC was introduced, neither Intel nor Microsoft recognized the opportunity. Intel was worried that its x86 architecture was uncompetitive and would soon lose out to the various RISC processors being developed by competitors like Motorola and National Semiconductor. And so, Intel embarked on its ambitious plan to develop the 32-bit 432. Microsoft, meanwhile, was working on an operating system, Xenix, which was based on Unix.

Compaq was founded in 1982 by three engineers working at Texas Instruments: Rod Canion, Jim Harris, and Bill Murto. Canion documented the story of Compaq in a 2013 book, *Open: How Compaq Ended IBM's PC Domination and Helped Invent Modern Computing.* Wanting to leave Texas Instruments and start their own company, Canion, Harris, and Murto explored several ideas, including developing an add-on hard disk for the IBM PC. Eventually, on January 8, 1982, they decided to create a portable version of the IBM PC. A nascent venture capital fund, Sevin Rosen, put together Compaq's first financing of $1.5 million, alongside the well-known venture firm Kleiner Perkins. Sevin Rosen had also, importantly, backed the start-up Lotus Software, which developed Lotus 1-2-3, the most critical application (a spreadsheet) for the IBM PC. A little over a year later, in 1983, Compaq began shipping their portable computer. They ended that year with an astonishing $111 million in sales.

Two initial strategic decisions set Compaq on a course for success. First, and most importantly, their computer was 100 percent compatible with the IBM PC. That meant that all software developed to run on the IBM PC would run on the Compaq computer, too. The second was to offer a portable computer, which was a product that IBM did not have.

To make their computer compatible, Compaq had to consider the BIOS (basic input/output system). A small sliver of software, the BIOS acted as the interface between the operating system and the computer hardware. It was often called firmware, because it did not change and was stored in read-only memory (ROM). Application software was supposed to access hardware through the operating system, not the BIOS directly. Still, application software developers would sometimes reach down into the BIOS to get higher performance from their applications. The *Apple v. Franklin* ruling in August 1983 (more about this in the next chapter) made clear that a BIOS could be copyrighted. Once IBM released its PC, many companies began to offer IBM-compatible PCs by just copying IBM's BIOS in the same way as Franklin had copied the Apple II BIOS. They therefore infringed on the IBM copyright in the same way as Franklin Computer had infringed on Apple's. Starting in late 1983, then, after the *Apple v. Franklin* ruling, IBM used the courts to put these other companies out of business. It was not pretty.

Compaq, by contrast, understood that they had to develop a fully compatible BIOS that did not infringe on IBM's. Compaq did so successfully, no easy task. Compaq also had to reengineer Microsoft's operating system, MS-DOS, so that it was compatible with IBM's PC DOS. Eventually, Compaq licensed its changes to MS-DOS back to Microsoft, as Microsoft had incorrectly assumed that application software for other personal computers would use its application programming interfaces (APIs) without reaching into the underlying hardware. The BIOS, Microsoft thought, would be unique for every computer manufacturer. As it turned out, that was not what happened. Clone manufacturers would routinely bypass the operating system to improve performance or reduce programs' size.

Compaq's second critical strategic decision was to offer a portable as their initial product. Since IBM had no portable computer, Compaq's portable provided customers permission and an opportunity, especially at larger corporations, to buy a non-IBM computer that was nevertheless software compatible with the IBM PC. By the time Compaq was formed, the market had some portable (luggable) computers, like the Osborne 1 and the Kaypro II, and these were doing well. Compaq believed that the combination of portability and compatibility would be a killer proposition on the market. It was!

CAN'T PUT THE GENIE BACK INTO THE BOTTLE

Compaq was the first to throw a sword into the body of IBM by designing a fully compatible computer. It was followed by tens of other companies. IBM never imagined that other companies could develop a compatible PC without infringing on their BIOS copyright, nor could they believe that such companies could succeed even with a non-infringing BIOS. In 1984, Lance Hansche, a senior executive at Phoenix Computers, discovered a failed clone company in Texas that had successfully developed a non-infringing BIOS. Phoenix acquired the BIOS and also made changes to MS-DOS, like Compaq did, offering full IBM PC compatibility to the many companies that wanted to offer clones without going through the expense and time needed to develop a legal BIOS. Phoenix's many customers included Hewlett-Packard, AT&T, Tandy, Gateway, and Dell. Now that any company could develop a computer that was fully IBM PC compatible, many did. By 1986, clone manufacturers had more sales of IBM-compatible PCs than IBM.

In 1984, IBM launched the IBM AT. This was perhaps the most consequential event in creating the PC industry. The AT utilized the Intel 286 processor, which was backward compatible with the 8088 that IBM used in its original PC. This solidified the importance of software compatibility, and unknowingly, IBM handed the keys to the PC kingdom to Intel and Microsoft.

IBM, soon enough realizing that it could not compete with the clone manufacturers that had a much lower cost structure, decided to develop a proprietary system, called the PS/2, with a proprietary operating system, called OS/2, and a new bus structure, called Micro Channel. They also used their proprietary networking technology, Token Ring. The PS/2 finally launched in 1987, and IBM paid Microsoft to help develop it. As silly as it may sound now, IBM paid Microsoft per line of code. So, of course, OS/2 was bloated, although it was probably technically rather strong.

IBM thought it could put the genie back in the bottle. PS/2 would be a success, and clone manufacturers would have to license technology from IBM. The resulting licensing fees would increase prices and provide IBM with the margins they needed to be profitable in the PC business.

At the same time, Microsoft was developing Windows. Gates tried to convince IBM to use Windows and abandon the development of OS/2. IBM refused and so found itself dependent on what would soon be their competitor in the operating system business: Microsoft.

IBM wanted to move from Intel microprocessors and use the PowerPC microprocessor, which they were developing together with Motorola and Apple. However, only Apple would use it in a personal computer. Compatibility forced IBM to continue to use Intel microprocessors. IBM based the PS/2 on the Intel 286 chip. Meanwhile, Intel was counting on the 386 to be a big success and tried to convince IBM to use it.

IBM strongly resisted using the Intel 386, the first 32-bit microprocessor that was compatible with the earlier x86 processors. They had concerns that a more powerful 32-bit PC could compete with their minicomputers, and they were exploring the development of their own 32-bit microprocessor with the PowerPC. In an interview with *PC Magazine* from March 25, 1997, Bill Gates stated that IBM was concerned that Intel was not capable of getting the 386 done. I doubt that was really IBM's view; most likely, it was just the excuse they gave Bill.

I'm still amazed that IBM thought they could succeed with this strategy. Perhaps if the PS/2 was vastly superior to the PC clone products, it might have had a chance. But it wasn't. It would have been much better for IBM to exit the PC business sooner—as it eventually did—to specialize in integrating PCs into the corporate world. After Asian PC companies entered the market, they drove prices down to the point that IBM had no choice but to exit the PC market altogether in 2004, wisely selling their PC business to Lenovo, a Chinese company.

Not having IBM's support for the 386 put Intel in a quandary. Intel decided to do everything it could to get Compaq to bring out a desktop PC using the 386, leapfrogging IBM. At one meeting at Intel I attended, someone said Compaq would be the rabbit that the dog (IBM) would chase. Once Compaq came out with a 386-based PC in 1987, the rest of the clone industry followed. The next year, Dick Boucher, an Intel vice president and my officemate, discussed the IBM situation with me. IBM had a license to manufacture a certain percentage of the 286 chips they were using. As Dick was the senior executive dealing with IBM, IBM had asked Dick if Intel would agree to let IBM manufacture more units, since they had unused capacity at their factory. IBM had negotiated rights to manufacture a portion of all future processors in the x86 family, including the 386. I suggested to Dick that Intel agree to let IBM build more 286 chips if they gave up the rights to the 386 and future x86 processors. IBM agreed.

IBM found itself in a difficult spot. The clone manufacturers did not move to IBM's proprietary technologies. IBM could not compete with the clone manufacturers, nor could they get the industry to move to any of their proprietary technology. IBM never brought the PowerPC to the desktop market successfully, though Apple did use the microprocessor for its Macintosh computers starting in 1994. The PC market that looked so good to IBM in the early 1980s was now a rock tied around their necks in a sea of clones, and IBM was drowning.

The American clone companies were not immune to the pressures of Asian competitors, either. Intel was very active in developing low-cost PC manufacturers in Asia that were willing to accept low margins. It served both Intel and Microsoft well to have many PC manufacturers, none with any market power. Effectively, the PC manufacturers became distributors for Intel and Microsoft products. Microsoft, in fact, could sell its products directly to end users, bypassing the PC manufacturers entirely, and even began to sell hardware peripherals like mice and keyboards, as well as application software like Microsoft Office. Access to end users was a very important advantage for Microsoft, one that Intel did not have. Intel's lack of such access is probably one of the reasons that Microsoft is worth so much more than Intel today.

Many years after the introduction of the IBM PC, Andy Grove described the change from a vertical to a horizontal computer industry in his 1995 book, *Only the Paranoid Survive.* He reflected on this change in the context of what he termed "a strategic inflection point":

> A strategic inflection point is a time in the life of a business when its fundamentals are about to change. That change can mean an opportunity to rise to new heights. But it may just as likely signal the beginning of the end.

The structural change in the computer industry from vertical to horizontal integration would take more than a decade; throughout the 1980s, the computer industry remained dominated by mainframes and minicomputers. The transition would only become clear to everyone much later. I think Andy thought the change would be permanent, but it did not hold through the ascendance of mobile computing devices, such as the iPhone.

6. GOING DIGITAL

Back in 1979, the impact the PC would have on the computer industry's future was still unknown. Having achieved recognition as an expert in the use of computer technology in cardiovascular medicine, I was going to give all that up to become a designer of computers. It was a scary decision. But at least I knew what company I wanted to work for. I had worked with Digital Equipment Corporation's computers for more than a dozen years, so that company was my first choice.

I got interviews with Digital Equipment Corporation through contacts I had developed over my long-standing relationship as their customer. Digital, at that time the second-largest computer company in the world, comprised many business units, called product lines, as well as some core services (manufacturing, engineering, sales, and finance). I interviewed with a number of the product lines and with a group run by Dick Clayton, whom I knew through Mort Ruderman. It was Clayton who organized and paid for my trip to Digital's headquarters in Maynard, Massachusetts.

A key person within Clayton's organization was Roy Moffa, and we hit it off. I wanted to work in his department. Within the dominion of the Digital Equipment Corporation, the engineer was king. To be successful there, I would have to prove that I was a great engineer, especially since I had no formal engineering training. I spent a few days meeting with Roy and other members of his team. Arianne, who had come to Massachusetts with me for this week of interviews, and I had dinner at Roy's house and met his wife. The week went by quickly, and Friday afternoon arrived before I knew it. Roy and I met for drinks at a restaurant near the Mill, Digital's headquarters, to wrap things up.

I fully expected Roy to offer me a job, and I was excited to accept it. Instead, he told me he could not offer me a position. While he thought that I could make a significant contribution to his group and the company, he felt strongly that the other managers in his group should agree. One of them did not, as it turned out. It was quite a blow. I was counting on getting this position at Digital; I didn't have a backup plan and had to take action. I emphatically told Roy that this was the wrong decision for Digital, for him, and for me. I asked him if I could speak with the manager who was not in favor of my joining the team, and I asked him to reconsider his position. The manager backed down, as I was pretty sure he would if confronted. Roy then hired me on the spot. I don't know what would have happened had he said no.

We flew back to Israel the next day. It took us about three months to arrange to leave Israel and move to the United States, which involved many complications, including how to get my share of the house sold. Fortunately, Moshe agreed to buy it for $100,000 in cash—and I mean bills. How he had so much in hard, printed US dollars was a mystery to me. Getting the money out of Israel, which had currency restrictions at the time, was a real challenge. I went to different banks with about $5,000 in cash, the maximum that could be transferred under the restrictions, and asked them to wire the money to my bank in the United States. Though terrified that the authorities would realize I was avoiding the currency restrictions, I was desperate to get my money out of Israel.

Arianne, who had both Dutch and Israeli citizenship, had to get permanent residence (a green card) to live in the United States. We also had to find a place to rent. I wrote to the synagogue in Sudbury, close to Digital's Maynard headquarters, and they found us a lovely home to rent for a year. It was exhausting, but we were there by July 1979. I remember the sense of relief we felt as we flew out of Israel. Now, some forty years later, I have a sense of relief when I arrive back in Israel, now my home once again.

Digital Equipment's headquarters, the Mill

Computer development within Digital's Central Engineering was organized into four quadrants: high-end and low-end hardware and high-end and low-end software. I became responsible for developing product strategy for low-end computer hardware. Then, just three months after I started, Roy moved to a new position, launching Digital's semiconductor capability in Hudson, Massachusetts. My new boss, Herb Shanzer, reorganized the group. I was now in charge of hardware development, product support, and strategy for low-end computers, which included all PDP-11s utilizing the Q-bus. It was a big job with more than a hundred engineers and technicians reporting to me. For the first time in my career, I did not know the names of all the people working for me.

In my new position, the manager who objected to hiring me ended up working for me. I fired him soon after, not out of any sort of revenge but because I thought that his opposition to my hiring wouldn't be the last time, as he put his insecurities above the well-being of the corporation. He quickly found another position at Digital.

While I did not work with Roy long, we stayed friends, and I will always be grateful for the opportunity he gave me. Roy died young, sadly. Many years later, ironically, Intel purchased the Hudson semi-conductor facility he had started.

Promoted twice within my first year at Digital, I must have been doing something right. I was in heaven, spending my time working with my team on product design and computer architecture. I had first discovered computers just twelve years earlier. Now I was running hardware engineering for low-end computers at a major computer company. Who could have imagined?

Herb and I got along well. We often ate Philly cheesesteaks at a cafe in Maynard and talked about the world. He was more friend and mentor than boss. My group was much larger than those I had managed in the past. Fortunately, Digital had internal courses and formal management processes, so I was able to develop some critical skills.

Ken Olsen, the legendary founder and CEO of Digital, was very focused on the mechanical packaging of computer systems. While he did not really grasp software, he had a strong interest in the physical design of computers, an attribute he would share with Steve Jobs. Ken had a poster printed and put it up all over the Mill with a photograph of the back of the company's DECmate word processor. Cables ran everywhere. At this point, I had just been promoted to be responsible for DECmate hardware, among other computer systems. While I thought that Ken's poster made a good point, I couldn't help myself. I scrawled on a number of the posters, "Can you imagine what software would look like if you could see it?" But I did agree that DECmate was a mess.

A COMPUTER FOR CLERKS AND CLERICS

Soon after, I received an invitation to attend the annual off-site meeting of about a hundred senior engineering staff. Outside speakers were invited. I most clearly recall a presentation by Al Shugart, the key person behind the Winchester disk technology still used in many personal computers. Al had led IBM's storage efforts, and he had just started a company, Shugart Associates. At that time, hard disks were large and very expensive. Al, realizing that the then nascent personal computers would become essential products of the computer industry, decided to make a small form factor consistent with their size, the same 5 ¼–inch size as the floppy disk units already shipping with personal computers. His initial product held five megabytes, just large enough to store one uncompressed photo taken by today's typical smartphone.

A few days later after, Gordon Bell asked me to attend a small meeting at the home of Stan Olsen, Ken Olsen's brother. Gordon Bell was and is still a very respected computer scientist/architect. He was the driving force behind most of Digital's computers and, most importantly, the VAX. Along with Stan and Gordon, two other members of the engineering organization, was Ken Olsen.

The meeting's primary purpose was to discuss the possibility of developing a small, single-user computer. But later, it became clear to me that it was really a job interview. In addition to engineering for low-end PDP-11s, my group was also responsible for the company's DECmate word processor and a small computer system offered by the commercial product line. I had already started designing a small computer system. Gordon had reviewed this work with me, so it made sense that I was invited to this meeting. At that time, a number of personal computers, like the Apple II and the TRS-80, had already hit the market, but Digital considered these products to be designed for hobbyists and unsuitable for business use.

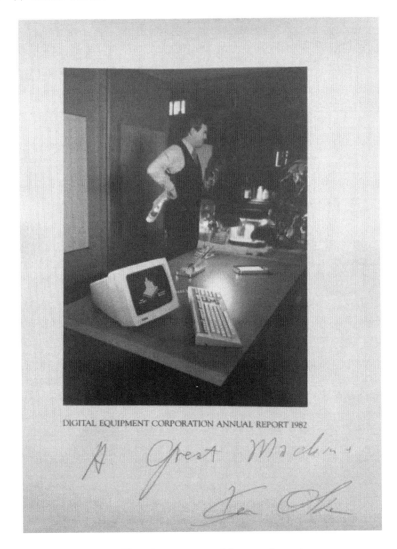

DIGITAL EQUIPMENT CORPORATION ANNUAL REPORT 1982

Ken Olsen gives me a copy of the annual report.

We never used the term "personal computer." I remember Ken saying that it should be so simple to use that even his minister and his secretary could use it. I jokingly referred to it later as a computer designed for clerks and clerics. Ken never considered that he might use a computer himself. Once when I asked him about that possibility, he told me his fingers were too big for the keyboard. But I knew that was not the reason. If he had really wanted to use a keyboard, he would have had one built with bigger keys.

Stan suggested that the computer monitor be tilted upward so that someone with bifocals could quickly look at the screen. It was a pretty good idea, one which we eventually implemented. I don't remember having a conversation about software during the meeting. Much of the decision that day had to do with the packaging of the computer.

After hours and hours discussing the features of this computer that did not exist, it became clear to me that no one in the room had any idea how to make such a product happen. I looked right at Ken and asked: "Do you really want this to happen?" He said he did. "Then I'll make it happen," I said. I was thirty-four years old. Ken was fifty-two years old, the same age as my mother. I thought then of Ken as close to retirement age, which is funny because now that seems so young to me. I do wonder where I found my apparently bottomless self-confidence at that age.

Ken questioned, during that meeting, whether it was actually possible to make such a computer at Digital. He even half joked that he and I should think about leaving Digital and starting a new company together. In retrospect, it is possible that everyone would have been better off if we had. I did not realize it at the time, and I guess Ken did not as well, but the company had grown too large for him to manage. Spending that day with Ken was a powerful experience for me. It would be the first time I spent time with someone of Ken's stature but far from the last.

For many years prior, Ken had not taken personal computers seriously. It is easy to understand. At the time of the meeting, the most successful personal computer was the Apple II, which had just a forty-character display. However, it did run VisiCalc, the first spreadsheet program, which was perhaps the catalyst for the PC revolution. Ironically, it had been developed on a Digital product—the PDT-11 Intelligent Terminal System.

Ken was very intuitive. It was that very intuition that led to the creation of Digital in the first place. Back in 1957, Ken had realized there was an opportunity to offer computers in what was then the low end of the computer market, computers that did not require air-conditioned rooms with specialized staff to operate them. He came to understand the potential of personal computers to create a new market but could not seem to get anyone in the company interested. At that time, the company's focus was VAX, the world's most successful minicomputer. The most talented engineers in the company under the leadership of Gordon Bell wanted to develop ever more powerful computers. I would observe a similar phenomenon later at Intel.

Ken was looking for someone to take charge and lead the company's efforts to create a personal computer. Gordon was looking for someone to get Ken off his back. This was to be me.

PROJECT KO

I told Ken that if he wanted this computer to happen, he had to give me the authority to do it. I asked him to invite me to the Operations Committee meeting (his staff meeting) scheduled a few days later. I explained that I would present a plan to the executives and would ask various groups to provide me with the resources I needed. I asked Gordon to write up our discussions and send out a memo to the members of the Operations Committee in advance, making it clear that Ken stood behind this project. I called the project KO, which I said stood for "knockout." Everyone knew it referred to Ken Olsen. We later renamed the program CT, for Communicating Terminal. I no longer remember why we took that name.

Gordon never had his heart in this project, focused as he was on the development of the VAX and VMS, Digital's 32-bit architecture. Meanwhile, we planned to use the 16-bit PDP-11 architecture. Gordon was glad that Ken had found something to keep him busy and away from VAX and VMS. Ken and Gordon clearly had some issues, to put it mildly.

The presentation to the Operations Committee went very well; even though I had never made a presentation to a group of senior executives before, I felt very comfortable doing so. The outcome was never in doubt because of strong support from Ken Olsen, Stan Olsen, and Gordon Bell. I asked each of the functional managers to provide me with someone to represent their organization, matrixed to me. I described my role as program manager for the KO project.

I no longer reported to Herb Shanzer. Elevated one level (promoted again), I reported directly to Dick Clayton, vice president of all low-end engineering. Dick was supposed to make sure that I got all the resources I needed, but he was pretty passive in reality. He also did not want to get in between Ken and me, so I met directly with Gordon and Ken to get things done. I must have complained to Ken about Dick's lack of interest, and as a result of this and several other issues, he was given a new role. Bill Avery replaced him. While Bill did provide the resources I needed, he did little else to help me but neither did he get in my way. I kept direct lines to both Ken and Gordon, frequently meeting with both individually and sometimes together.

The first person recruited to join my team was Vahram Erdekian, an engineering manager in the manufacturing organization. He was outstanding and made significant contributions to our effort. Ron Ham joined us as the head of software development. Art Williams was responsible for hardware development. Mike Weinstein joined the team to run our marketing, and Bob Sanfacon ran operations. For a while, I even made Ken Olsen the packaging and power supply engineer, and eventually put him in charge of all mechanical engineering for a time. The actual engineering was done by Dick Gonzales, the senior mechanical engineer at Digital and a man who was very close to Ken. But Ken did sign off on Dick's designs. I knew this area interested him the most, and it kept him engaged and committed to the program. It also gave me a reason to meet frequently with him alone. I realized that even though he was the CEO of a major corporation, he missed being a design engineer.

The team, except for Ken, met every day both to define our product offering and to figure out how to get everything done. It was exhilarating, and I was very happy. I could not believe my good fortune. The next two and a half years would be the most intense of my career, and I learned so much about leadership. I had found my voice.

Ken wanted a computer that was beautiful to see and simple to use. His limited experience using computers was word processing, and his assistant did that for him. I doubt that he ever touched a keyboard himself. His vision was a computer that was powerful, easy to use, beautiful, and at a low enough price point that every professional and small business could afford one. In many ways, Steve Jobs would later remind me of Ken. Ken saw this as a new market, as larger enterprises and institutions were already served by time-sharing computers accessed via terminals. In 1980, Digital's core business comprised of the VAX and PDP-10, many of which operated as time-sharing systems. Indeed, it was on a PDP-10 time-share computer where a young Bill Gates would learn to program. Enterprises, even the personal computer companies such as Apple itself, did not use personal computers to run their businesses. More than a decade would elapse before that began to happen.

We defined a system called the CT325 to meet Ken's design requirements. My heart was not in that product. It was very limited, with small memory (128 kilobytes), and only a floppy disk for external storage. It was undoubtedly beautiful, however. We designed a much more robust system, the CT350, meant as a workstation for professionals. This is where we directed much of our effort and budget. Later, we changed the name from CT to the Professional to reflect our focus. The DEC Professional 350 was elegant, powerful, and, ultimately, a total failure. I still love it.

Ron Ham developed our software architecture. Digital was slowly moving to a single computer architecture under Gordon Bell's technology leadership. The hardware was VAX, the software was VMS, and the network was DECnet running on top of Ethernet, which had just been designed. Digital was already making significant investments in semiconductor capability through the facility in Hudson run by Roy Moffa (and Steve Thiecher). Their target was VAX on a chip, but that would be many years in the future. We want to make the Pro 350 compatible with VMS, the VAX's OS, at the application layer. That way customers could migrate to the VAX architecture when we eventually had a VAX/VMS version of the Pro.

To accomplish that, Ron advocated that we use an operating system called RSX-11, an exceptional multitasking and real-time operating system developed for the PDP-11. It would be a decade before Microsoft would be able to offer such capabilities. Applications were written for VMS in the BLISS programming language, which was created at Carnegie Mellon University. As the primary programming language used for VMS application development, Ron figured out how we could recompile those applications on RSX-11 to offer a great many applications that were also offered on VAX/VMS systems. The Pro 325 ran a low-end operating system, RT-11, and was not software compatible with the rest of the Pro family. Had the IBM PC never launched, resulting in an eventual restructuring of the computer industry, this strategy might have been very successful. When it was proposed, we had no knowledge that the IBM PC was in development, which was kept very secret by IBM.

My group had about three hundred people directly, alongside probably more than one thousand people in other groups including manufacturing, service, marketing, and sales. Our initial budget was $20 million dollars. There were only thirteen people within IBM's Boca Rotan's Entry Systems Group developing what became the IBM PC.

Gordon had no interest in low-end computers and kept himself busy developing ever more powerful VAX systems. He did have a passing interest in workstations, and we visited Sun Microsystems together around 1982. We briefly discussed trying to acquire Sun, but Gordon was not interested. Digital really should have owned the workstation business that Sun and later Silicon Graphics dominated. It would have been a natural fit for Digital, but I didn't press the issue; my heart was really in developing systems for professionals (knowledge workers).

THE WORLD'S MOST POWERFUL PERSONAL COMPUTER

Several critical design decisions defined the Pro. We decided to develop a unique dual-sided, five-inch floppy disk drive. Floppy disks at that time were single-sided, so we doubled the amount of storage while keeping the same size enclosure. This was an example of an engineering-driven decision that resulted in our floppy disks not being compatible with any other computer system. The display was bitmapped, and it used the same memory space as the programs and data, which was a very advanced concept for its time. Wim Englse, whom I brought over from Rotterdam, and I developed this over the objections of the Pro engineering team, who thought that it would slow down performance too much and did not appreciate the power it could provide. Wim and I had a lot of experience with bitmap graphics from our work much earlier at the Thoraxcenter. Using a bitmapped, directly addressable display enabled powerful formatting on the screen, including advanced fonts. In 1981, the Xerox Star was the first commercial system to offer bitmapped graphics on an eighty-character display, but it was priced about five times higher than the Pro. Wim and I won an award for our design of the graphics unit. I was and still am very proud of it.

Personal computers (PCs) at the time were desktop devices. Since we did not want to take up valuable desktop space with the computer enclosure, we designed the Pro so it could not only lie flat but also be rotated ninety degrees to stand vertically on its side. This way, the monitor and keyboard took the bulk of the valuable desktop space. Later, other manufacturers would adopt this design, and it would become the dominant way of orienting desktop computers.

Professional 350

Working as a consortium, Xerox, Intel, and Digital agreed to develop and launch Ethernet, which was a way to connect computers together if they were located physically close to each other. The first specification was published in 1980, so Gordon asked me to incorporate Ethernet into the Pro. We added a board from 3Com, the company founded by Bob Metcalfe, who had developed the original concept of Ethernet while at Xerox. The initial design of Ethernet used a coaxial cable, similar to the ones used for television. For some reason, it was yellow. If you wanted to connect the Ethernet, you used a clamp that not only gripped the cable but also penetrated it, creating a physical connection. One day, when I was in Gordon's office in the Mill, Ken walked in with a yellow cable in one hand and a bunch of phone cables in the other. Ken said to Gordon, "Maybe your university friends want these yellow cables, but my friends want telephone wire." Gordon, angrily, snapped back, "Why don't you go work at Data General and do an Ethernet with phone wire. I hear they are looking for engineers." Ken was right; telephone wiring ended up as the standard. Interestingly, neither Xerox, Digital, nor Intel directly benefited from their work defining Ethernet. New companies, like 3Com, and later Broadcom, would take advantage of the opportunity they had created instead.

The key to our system was a five-megabyte, 5 ½–inch Winchester hard disk that I had seen at the Central Engineering retreat. Finis Conner was head of sales of Al Shugart's company, which was now named Seagate, the only supplier at that time to offer a hard drive in this small form factor. Finis asked me to give him a $1 million purchase order in advance of the Seagate IPO. I did, which I'm sure was beneficial for them. Had I been more sophisticated at the time, I might have asked for some warrants so that Digital would have made some money on the increase in their value as a result of our order. I almost certainly did not even know what a warrant was at that time. We had a lot of problems with their first shipments. I used to joke that Al Shugart taught me that you could sell, manufacture, and then design a product in that order.

Digital never became a large customer for Seagate, because we were unsuccessful in the personal computer market. Their primary customer became IBM, with the IBM PC XT. Meanwhile, the disk engineering group at Digital, run by Grant Saviers, was actually in the process of developing a five-megabyte Winchester disk themselves. Before I could order any Winchester disks from a competitor, I had to convince the disk engineering group to approve this plan. Expecting delays, I didn't want to depend on an internal engineering group for critical parts. I suggested to Grant that his group develop a ten-megabyte hard drive so they could leap ahead of Seagate. I told them I was confident they could launch such a large drive soon after Seagate brought their smaller five-megabyte drive to market and promised to switch to the Digital drive as soon as it was available. When Seagate launched a ten-megabyte drive before the internal group could, I had to convince them once again, this time to develop a twenty-megabyte hard drive. I don't know if they ever got that done, either, because I was gone from Digital before it was supposed to happen.

The keyboard, as it turned out, was one of the most challenging components to develop. The technical challenges were modest at best, but the critical business units couldn't agree on the labeling of the keys. We set up a keyboard committee comprised of the various interested parties that met every week and never came to an agreement. The keyboard was named the LK200. Finally, fed up, I went to the committee meeting and told them how proud I was of their work and dedication. I explained they could take as much time as they needed. In the meantime, I said, we should start working on the successor keyboard, to be called the LK300. Since they were all so busy with the LK200, I suggested, I would find other people to work on the LK300. That week, the committee finished their work. All the members then volunteered for the LK300 committee, which I never formed. I was so glad to be rid of them.

In 1980, the telephone and handwritten or typed letters either mailed or faxed were the primary forms of personal communication. Voice-mail systems from companies like ROLM, Octel, and AT&T were just being introduced. I decided that we should integrate telephony capabilities into the Pro. We called it the Telephone Management System (TMS). The idea was to provide a professional voice-mail option in a small, remote office as if it were a central office. The user could select from a contact list to dial and record voice messages. There was a built-in 300-baud modem, the maximum speed for a modem without a direct connection to the phone line. With a direct connection, one could go as fast as 1,200 baud. By comparison, connections to our homes using cable modems today achieve speeds between a thousand and ten thousand times faster.

We worked together with VisiCorp, distributors of the hugely successful spreadsheet, VisiCalc, developed by Software Arts. VisiCorp distributed several other applications in addition to VisiCalc, with only the "Vis" at the beginning of their names shared in common. The company decided to create a product, Visi On, that integrated all of these applications, modeled after the Xerox Star. The Pro 350 was their target platform, as it was the only personal computer with a hard-drive operating system and enough memory and power to support this ambitious software project. VisiCorp was larger than Microsoft when Bill Gates saw Visi On demonstrated at the 1982 Comdex trade show in Las Vegas on a Pro 350. The documentary *DEC: Personal Challenge, 25th Anniversary Video*, produced by Marc Porat, focused on the development of the Professional. It shows Ron Ham, Mike Weinstein, and me walking around the 1982 National Computer Conference in Houston, clearly thinking that we had just won the PC war. The CEO of VisiCorp, Terry Opdendyk, and Bill Gates had serious discussions about the two companies merging, but these talks collapsed. Seeing the Visi On demo probably influenced Gates's decision to build Windows.

We wanted to develop an independent application ecosystem, so we devised the idea of "digital authorized software" with three levels: platinum, gold, and silver. Platinum software was marketed, distributed, and supported by Digital. Gold software was certified by Digital, meaning that we tested the software in our labs. Silver software was self-certified based on a testing process we had developed. We signed up a lot of software companies, particularly in vertical areas like finance, health care, and education, and those companies paid Digital for certification. The product lines drove the development of the application business.

Periodically, I gave updates on the project to the Operations Committee. I was also asked to update Digital Equipment's board of directors, which was pretty heady stuff for me. The board at that time included Georges Doriot, father of the modern venture capital industry; Philip Caldwell, the CEO of Ford Motor Company; and, of course, Ken Olsen, who later joined Ford's board.

The Operations Committee asked me to rehearse my presentation with them before I gave it to the board. This was a problem; unlike the Operations Committee, the board was not very technical, so the same presentation would not really go over well for both. I carried a bag of cables and power supplies into the Operations Committee, set them down on the table, and said, "You can really help me if you can do a bit of role-playing and act like the people on the board instead of executives of a computer company. But, if you would rather do the latter, I brought some hardware with me." Somehow, this stunt worked. The presentation to the board went well, although the board had no clue what I was saying. I think they were just trying to figure out if I knew what I was doing, a test I apparently passed. As I told the Digital board, the Professional represented Digital's initial attempt to address the PC market by offering a proprietary system that was compatible with Digital's PDP-11 line of computers.

Coincidently, around the same time I was laying out my plans for the Professional Series PC to the Management Committee of Digital Equipment, Bill Lowe was presenting to the IBM Management Committee in August 1980. While IBM had just thirteen people working on its PC and would ship it twelve months after the start of the project, I had hundreds of people working on the Professional at Digital and spent three years developing an unsuccessful product. The Professional would not ship until mid-1983, by which time IBM and manufacturers of IBM-compatible computers would be well entrenched in the market. The IBM team traded off functionality and quality in favor of time; the one-year deadline was held absolutely constant. At Digital, we instead favored functionality and quality. We failed to understand what Adam Osborne, the founder of Osborne Computer Company, meant when he spoke of being "adequate." While the IBM PC Group primarily bought off-the-shelf components, Digital designed almost everything from scratch.

HENRY FORD'S OFFICE

Digital created a system of executive sponsors, executives who were assigned to look out for different companies. As an executive sponsor, one had to make sure the customer was happy and give them someone to call if there were serious issues. Ken asked me to be the executive sponsor for Ford Motor Company. I considered this a great honor, since Ford CEO Caldwell was on the board of Digital and Ken was on the board of Ford.

Toward the end of 1980, Ken asked me to go to Ford's headquarters in Dearborn, Michigan, to give Caldwell's assistant a DECmate computer. The Digital salesperson responsible for the Ford account picked me up from my hotel in Dearborn the morning after I arrived. He drove a Ford and explained that all people who dealt with Ford had to drive Ford cars. We got into the parking area, and indeed, I saw only Ford cars. I asked him if Caldwell drove a Ford. He said of course he did, but he did not drive it himself, and the vehicle was serviced every day.

We took an elevator dedicated solely to the ten or so senior executives on the floor where Caldwell had his office. The boardroom was one floor higher. This part of Ford headquarters had rooms where board members could actually stay overnight. The walls were marble. Each executive had an executive suite with a waiting room, an office, and a place for their assistant, back then called a secretary. I was introduced to Caldwell's assistant. She first showed me Caldwell's office, which had been the office of the founder, Henry Ford. It was filled with millions of dollars' worth of art.

We then went into the assistant's office. I explained to her how the DECmate word processor worked. She looked at me, confused. Eventually, she asked, "Why would I use this?" I replied it would help her with her typing. She laughed. "I do not type anything. If something needs to be typed," she said, "I send it down there." She pointed at the floor. "What's down there?" I asked. "The typing pool," she explained, perhaps a bit irritated. I told Ken she loved the gift, hoping he would never find out that she would never use it.

AND THE COMPETITION WAS . . . US

In August of 1981, IBM released its model 5150. I got my hands on the IBM PC soon after its release. I took the IBM PC into my lab at the Mill, and invited CEO Ken Olsen to join me there. Ken had a background in power supplies and mechanical engineering. He asked me for a screwdriver, and we began to take the IBM PC apart. I don't think we ever actually turned it on. Ken was disgusted at its construction. He looked me right in the eyes and said that if I ever built anything like this, I would no longer be welcome at Digital. He never asked a single question about the PC's software, which was not so surprising. Ken famously said that if you build great hardware, the software will come from heaven.

I agreed with Ken; the IBM PC was not impressive. I could not figure out why anyone would buy it. About a year later, Digital decided to develop an IBM PC–like computer, the Rainbow. It was not fully compatible with the IBM PC; the requirement that application programs be modified is one of the main reasons it failed in the market. Of course, as we now know, it was a start-up, Compaq, that understood the need to be 100 percent compatible. They introduced their first product in 1983 and ended that year with an astonishing $111 million in sales.

But back in the summer of 1981, I had been working hard on the development of the Professional for more than a year. I visited Digital's European headquarters in Geneva to engage that management team to help develop the European rollout for the Professional. Before that meeting, I took a few days' vacation and went to the Algarve in Portugal. There, I got a nasty cold. By the time I got to Geneva, I was very ill and had to cancel all of my meetings and fly back to the United States.

When I arrived back at my office at the Mill, I discovered that Barry Folsom, the engineering manager of the Components and Terminal Group run by Andy Knowles, had in my absence made a proposal to Ken Olsen. He said that his engineering team could build a product that would compete directly with the IBM PC. Barry had already developed the VT180, called the Robin internally. It was a CP/M add-on board to one of the terminals the Components Group sold. Ken was still longing for his low-end computer, the one I nicknamed "the computer for clerks and clerics." Barry's proposal was ingenious. He would use the packaging, keyboard, monitor, and other peripherals that we had developed for the Professional, but the motherboard would be different. It would include two CPUs: a Zilog Z80 to run CP/M, and an Intel 8088 (like the IBM PC) to run MS-DOS. Barry said he could get the product done in less than a year and at a price point consistent with Ken's objectives. Since it looked just like the Professional, it would also have all the beautiful packaging that Ken loved. At the same time, Dick Loveland, product manager for the DECmate word processor, said he could do the same thing with a PDP-8 chip, offering a version of the DECmate that would also use the same packaging as the Professional. Ken took these proposals to the Operations Committee for a rubber stamp. All of this happened in the very short time I was away.

I was furious, but I didn't know what to do. I was convinced that Barry knew he could not make his targets. I found myself not only trying to develop and launch the Professional but also managing much of the engineering, manufacturing, and service for two additional products: the Rainbow (Barry's product) and the DECmate II. This would greatly complicate things for the production, service, and sales organizations.

Software was the real killer. Incredibly, Barry and I were actually out in the world competing against each other for software. Barry was very much aligned with the Microsoft world and even brought Bill Gates out to meet with Ken Olsen. I attended that meeting, and it was the first time I met Bill, who was just twenty-four years old at the time. I remember little about the DECmate II project other than that I could no longer get the DECmate engineers to provide word-processing software for the Professional, because they were so committed to the DECmate II. This meant we could not offer Digital's proprietary word-processing software on what was planned as the company's flagship PC product.

Meanwhile, an adamant Bill Gates convinced Barry that as long as our hardware ran MS-DOS, the Rainbow's application compatibility with the IBM PC would be tied to the MS-DOS application programming interfaces. It never occurred to any of us to build the DEC Rainbow into a fully IBM-compatible PC like Compaq would do. Gates believed that software compatibility would happen at the level of MS-DOS, and he convinced Folsom of this. Gates was completely wrong about this. To get performance out of the PC's cheap hardware, software developers started accessing the underlying hardware, especially for graphics. Compaq and Franklin would soon prove the importance and promise of full hardware-level software compatibility.

Ken, of course, understood these software issues not at all. The company's sales force was greatly concerned about having three products that looked exactly the same but that ran very different software and addressed very different markets. We were confusing both third-party software developers and the various distribution channels. It was awful!

Eventually, Barry, Dick, and I all had to begrudgingly agree that introducing three products would lead to all three failing. We asked Andy Knowles, who had the business responsibility for our products, to organize a meeting between all of us and Ken. At that meeting, we expressed our concerns and recommended that we bring only one of the products to the market. Ken would not have it, saying famously, "Let the customers decide." In response, as I was later quoted in Glenn Rifkin and George Harrar's 1988 book, *The Ultimate Entrepreneur*, I said: ". . . and they did. They bought IBM."

Ken Olsen was exceptionally charismatic one-on-one. He would sit or stand very close to you and speak slowly, like there were only two people in the world, you and him. It was very seductive; you felt like you would do anything for him. I have heard that Steve Jobs had a similar effect on people. Though I met Jobs several times, I never met him in a one-on-one setting, so I can't confirm this. Ken called me often on Sunday evenings, always speaking in his slow, seductive voice. "I know you are very busy," he would say to me, "but I am worried about Joel's ability to do his job." He would ask me to keep tabs on Joel Schwartz, who reported to Andy Knowles and who was responsible for selling the Professional and the Rainbow. Often, after hanging up, he would call Joel at home and say the same thing to him about me. Joel and I would then call each other and compare notes. I found out from others at Digital that this behavior was not that unusual. Ken liked to pit people against each other just like he wanted many product options.

Once, Ken told me how difficult it was to keep a company like Digital successful. He said, "There are a thousand new companies trying to take away parts of our business. Most will fail, but some will be successful. This means we have to be successful with whatever we do. We can't fail." This is really a critical issue for most successful companies. Ken's words would come back to me later at Intel.

Meeting with Ken in his office at one point, we discussed the importance of marketing. I said that for Digital to be victorious in the PC space, we needed great marketing and good products. Ken disagreed. He said we needed great products and good marketing. We went back and forth.

Eventually, losing patience with me, Ken said forcefully, "You don't understand! We are not capable of doing great marketing." At that moment, I realized that my destiny was not to stay at Digital. I began to think about what my next steps should be. I realized that products like the Pro would have to be marketed to the actual users of the product and not just to the IT departments. That would require a different kind of marketing, but I failed to make my case. I would see the same circumstance again later when I was an executive at Intel, but in that case, Intel made the transition.

The day before a meeting of several hundred senior executives, Ken called me into his office. "People think that I treat you special," he told me. "While that may be true," he continued, "I don't think it's good they feel this way. So, tomorrow at the management meeting, I am going to criticize you. I just wanted to let you know in advance. I still believe in you." I was not surprised, and there wasn't much I could do. I just thanked him for letting me know in advance. Ken was more than a bit Machiavellian.

Sure enough, during the meeting, Ken said, "Did you know that Avram Miller wants to do marketing? Why would someone capable of doing excellent engineering wish to do marketing?" Everyone laughed. I thought not only about how his remarks might affect me personally but also about how they might affect the marketing people who were attending that meeting. In my mind, it was just one more example of Ken not understanding the importance of marketing. He was right; Digital was not capable of great marketing. What kind of marketing people would want to work for a CEO who saw marketing as such a second-class role?

Note the screwdriver in Ken Olsen's hand.

DIGITAL EQUIPMENT FAILS THE PC TEST

In May 1982, we launched all three products. Right after the announcement, I flew on one of Digital's corporate jets to Palm Springs to attend the PC Forum, organized by Ben Rosen, who later became the chairman of Compaq Computer. That night, I had dinner with all the speakers, including Bill Gates, Steve Jobs, Adam Osborne, and Mitch Kapor. Gates and Jobs were both twenty-six years old. I was thirty-six years old. Sitting in that room were people who would literally change the world. I was one of them.

At the meeting the next day, I gave a presentation on Digital's personal computer initiative. As a representative of the second-largest computer company in the world, there was a lot of interest in what I had to say. At some point during the other presentations, I found myself sitting next to Steve Jobs. We had a few conversations between speakers. I sensed that Steve was considering asking me to join Apple. He did not. I would not have accepted. At that point, I still believed the

Professional would be a great success. While this was the first time I met Jobs, it would be far from the last. I had no idea about the profound impact he would have on my life.

At the conference, Ben took me aside to show me the prototype Compaq Portable, which would be released in March 1983. It made little impression on me. I found it difficult to imagine people carrying around this "luggable," which weighed twenty-eight pounds, looked like a sewing machine, and cost more than $3,000 (more than $7,000 in today's dollars). I did not understand the coming revolution in the computer industry, and it would have been incomprehensible had you told me that, just sixteen years later, Compaq would acquire Digital.

7. RIGHT BIRD, WRONG NEST

By the second half of 1982, it was evident to me that Digital's personal computer efforts would fail in the market. By that time, the IBM PC was a clear success. Compaq had introduced its portable. Not only did we have strong external competition, but the Pro would have to compete with two internal products, Rainbow and DECmate. There was nothing I could do to change this outcome, nor could I see a long-term role for myself at the company.

Deciding to leave Digital, I considered starting my own company focused on computer networking, which I saw as a big opportunity. 3Com had formed in 1979 to sell Ethernet systems. Cisco was not started until 1984. IBM was pushing a different standard, called Token Ring, which would not arrive until 1984 and would not be a commercial success. So, 1982 was a good time to be thinking about computer networking. I was particularly interested in how personal computers might be connected.

Several things kept me from striking out on my own. I had no idea how to raise money, nor did I have any of my own. It's funny to realize that less than ten years later, I would become one of the most active investors in early-stage high-technology companies in the world as a co-founder of Intel Capital. I knew that the Professional would not be a success, and start-ups are very risky, and I was scared to fail twice in a row. I quietly discussed the idea with a few of the key people on my team at Digital. I felt sure that some would join me if I did start a company. I was unsure what to do.

That is when I got a call from a headhunter working for Franklin Computer, a company making an Apple II clone that we once laughed about in an Operations Committee meeting at Digital. The idea that a start-up could succeed by selling a copy of a leader's product was incomprehensible to Digital's executives. Just fifteen years later, Digital would be acquired by Compaq, a company that had started out as an IBM clone.

In February 1983, when I first began talking with Franklin, it was doing very well. Its first-year revenue was $28 million, and that would increase to $75 million a year later. Apple had sued Franklin and lost in the lower court and was appealing the judgment. In addition to Franklin's strong sales, it was building a large dealer network. Many retail computer stores wanted to sell the Apple II, but Apple was very restricted in its distribution. Franklin would have about $80 million in its first full year of sales and was on a growth track similar to Compaq's.

Franklin's board had some heavy hitters. The chairman of the board, Jim Simons, was a real gentleman; I liked him very much. He was also extremely smart. Demonstrating those smarts, he later became one of the most successful hedge fund managers ever, sometimes personally earning more than a billion dollars per year. Jim started life as a mathematics professor, chairing the math department at SUNY. Together with Shiing-Shen Chern, Simons developed the Chern–Simons form that has played an important role in string theory and quantum mechanics. He made his first fortune trading currencies.

A second board member was Howard Morgan, professor of business at the Wharton School and a partner in a hedge fund, Renaissance Capital, with Jim Simons. Renaissance Capital became successful, and in the process, Simons made tens of billions of dollars. Howard eventually became a venture capitalist and co-founded First Round Venture. The three founders of the company were also on the board: Barry Borden, Joel Shusterman, and Russ Bower. With the exception of Russ, all the board members were Jewish. That made me feel more comfortable with taking the risk. I agreed to meet with the company. Howard played the lead role in recruiting me.

The founders were very entrepreneurial. Both Barry and Joel had retail computer experience, while Russ had some engineering capabilities. While they had made some significant improvements to the Apple II, Steve Jobs was not that far off the mark when he claimed that Franklin had "xeroxed" it. The team at Franklin realized that 100 percent comparability with the Apple II was a critical requirement for success.

Given Franklin's sales momentum and Apple's loss in court, nothing, it seemed, could stop the company from going public. There was one exception: none of the people with the company were qualified to run it. That's where I came in. They wanted me to put together a complete team to run the company. My title would be chief operating officer and, later, president, but in all practical ways, I was the CEO.

We discussed compensation, and Franklin closed hard. They offered me $1 million in stock, vesting over four years. These were not options but a grant. I would not pay for the stock, and the company agreed to pay the tax. In today's dollars, that amount would be about $3 million. Of course, if the company went public, the shares could easily be worth ten times or more.

I said yes. The opportunity to shape and mold Franklin, the reduced risk of joining a company at this stage compared to starting my own, and the substantial compensation all played a role in my decision.

Before I gave my final answer, I asked the company's outside attorneys about the Apple appeal. They assured me Apple had no chance to win. So now I was a millionaire on paper, but just a year later, I would be out of work and broke.

TIME TO MOVE ON

Back at Digital, I met with Bill Avery, who was still technically my boss. I told him I had decided to resign. He asked me to stay. I next heard from Jack Smith, who was, in effect, number two at the company. He asked me to come to his office. We discussed my decision, and he asked me to reconsider. I told him I would think about it, but I doubted I would change my mind. An hour later, Ken Olsen asked me to meet him in his office. Talking with Ken was difficult for me emotionally. He was very much a father figure. Although I had a lot of criticism of Ken, I was very fond of him personally. I would not talk with Ken ever again.

Ken was very kind to me during our conversation and said he could understand why I would want to go off to a much smaller company. He might even have been a bit envious. Ken then asked me to do something to which I should not have agreed: he asked me not to take anyone from Digital with me. I told him that were two people I wanted to take with me. First, Vah Erdekian had said to me in the past that he wanted to leave. I had asked Vah to wait and join me when I left, and he had agreed. I explained this to Ken and said that I wanted to honor my commitment to Vah. The other person was Wim Englse, whom I had brought from Rotterdam. I felt that I had to offer him the opportunity to join me as well. I had so much respect for Wim that I hoped he would, but he wisely stayed at Digital. Wim eventually started his own company and did exceptionally well. I sometimes see photos of him on Facebook on his sailboat. I did not realize at the time, the handicap I had accepted by limiting the people I could have hired from Digital.

BANNED IN MAYNARD

One thing I needed to do before resigning was to contact Marc Porat. Digital had contracted him to make a film, *Personal Challenge*, for Digital Equipment to celebrate the company's twenty-fifth anniversary. The film, focused primarily on the development of the Professional, was scheduled to be released soon. Marc had spent more than a year with me and my group. I told him in the summer of 1982 that I would likely leave Digital. That is why the last scene in the film, taken right after we had demonstrated the Professional at a major trade show in Houston, has members of the team, including me, dancing at Gilley's (a club we knew from the 1980 John Travolta hit, *Urban Cowboy*) to a cover of a Kenny Rogers song:

> *Let's go out in a blaze of glory*
> *All good things must end*
> *Like two heroes in a story*

Let's go out like we came in
In a blaze of glory

Soon after I resigned, my staff came over to my house to view the film. I had copies made, which I gave to each of them knowing that Digital would not allow it to be distributed after I left. Gordon Bell got a copy, as well, which he eventually gave to the Computer History Museum. The museum subsequently has uploaded it to YouTube. Though the video is a bit embarrassing to me now, I am nevertheless glad it is out there as a historical document of our work on the Professional series.

Marc would later join Apple Computer. He then created the infamous company General Magic, started within Apple and spun out in 1990. General Magic was staffed by some of the best people in the computer industry and backed by leading companies including Sony, Motorola, Philips, and AT&T. It attempted to develop the world's first personal digital assistant and failed. Like our failure at Digital with personal computers, it was another example of trying to develop everything at the same time. A 2018 commercial documentary on the company, *General Magic*, is definitely worth watching.

Many senior executives left Digital in 1983, including Gordon Bell. The *New York Times* ran a long article on this on September 4, 1983, under the headline: "The Reprogramming of Digital." I am quoted, but the most interesting and insightful quote was from Ken:

> I've been criticized for holding on to my friends too long . . . Some were so rich they didn't want to work hard, and some want to become so rich. A gentleman never explains why someone left, but I'm happy with the people we have today . . . My plan is to stay here as long as I stay healthy, and you don't indicate an heir apparent when you have a president who's healthy and who has a number of years left.

Ken was saying that he was fine with so many key people leaving. They would not be missed; he would not need them. He could run the company without them. This attitude would be repeated later at Intel, when Paul Otellini, then CEO, would tell me that he would not trade any of the executives at Intel for the ones that left (including me, I presume).

Apple II next to Franklin Ace

FRANKLIN ON A ROLL

Soon after I joined Franklin, we approached Apple and offered to pay a royalty for every system we sold. We tried to set things right despite the lawsuit. We said we would expand the market. Apple CEO John Sculley seemed interested. Steve Jobs, though, said he wanted to crush us. True to his word, he did. Amazingly, just ten years later, I would be sitting with him at his kitchen table in his Emerson Street, Palo Alto, home like none of this ever happened.

Franklin planned its IPO for early 1984. Montgomery Securities and Smith Barney were selected as the underwriters before I joined, with talk of valuation around $350 million. The prospect of such a lucrative IPO was a significant factor in my decision to join. My job was to use the company's sales momentum and the money raised through the IPO to build a significant computer company, not just a company churning out Apple II clones.

— take your team if you want to have a chance

To accomplish this, I would need to build an outstanding team. Vahram Erdekian, who worked for me at Digital, agreed to join me as VP of manufacturing. But I honored my promise to Ken Olsen not to poach others from Digital. That was a big mistake. I decided to hire the best people from different companies for each of the significant functions. I did not realize at the time that this would mean that they would not share a common culture. Franklin itself was too young to have established a culture. I did not yet really appreciate the importance of one.

Our headhunter told me that it was possible to hire Bill Syndes, the project manager for the original IBM PC. He was responsible for the development of the PC Jr. at IBM, a consumer version of the IBM PC. That product, which turned out to be a major disaster, did not ship until December 1983. In April 1983, before we knew of the PC Jr.'s looming failure, I met with Bill at Franklin's headquarters in Cherry Hill, New Jersey. Everyone on Franklin's board was very excited to imagine that a critical person behind the development of the hugely successful IBM PC could become Franklin's VP of product development. I was, too, until I met Bill. To put it simply, I was unimpressed and could not imagine that he had made the IBM PC happen, but he had. As a technical/product guy myself, I wanted a VP of development with whom I could work closely, which was not Bill. I told the board my concerns. The board just could not imagine passing on Bill and put a lot of pressure on me. They said we just had to hire him. Once we completed the IPO, they said, "If you're still not happy with him, you can let him go." Giving in to the pressure, I made the offer, and Bill accepted. Bill wanted to bring on Lew Eggebrecht, who had been the architect of the IBM PC. IBM had later fired him, alleging that Lew had misused IBM's proprietary information. Lew was a very talented engineer and a substantial addition to our team.

I hired Gene Sherman as head of marketing and sales. He had been VP of sales at Exxon Office Systems and brought on Sheldon Chizever from Coopers and Lybrand as CFO. The board thought I had assembled a dream team. Unfortunately, it was anything but. Bill and Vah couldn't get along, with different ways of thinking about product development, manufacturing, and quality assurance. There was a constant rift between engineering and manufacturing.

Once the prospectus for the IPO was completed, we started our road show, the first time I would meet with potential investors. We gave a presentation at the famous Carlyle Hotel in New York. The market for PC companies was starting to get hot. In December 1983, Compaq went public, raising $67 million. I vividly remember the Franklin outside attorney, Steve Greenberg, telling potential investors that there was no way Apple would prevail in the appellate court.

If the bankers were correct about the valuation, my shares would be worth about $10 million ($25 million in 2020 dollars), which seemed like an incredible amount of money, especially at that time in my life. I told myself, wisely, not to count on any of it.

— don't believe predictions from lawyers

BEST-LAID PLANS

We were just waiting for the appellate court's decision before taking the company public. Everyone was so confident that the court would not reverse the lower court's decision that I didn't even bother to attend the court session on the day it released its decision. Barry Borden called me later and explained that based on "matters of law," not "matters of fact," the court decided the case should be heard again in the lower court. I was stunned. Franklin's legal team had presented many arguments to the lower court, one of which was that microcode or firmware could not be copyrighted. The appellate court believed it could be. Franklin had taken the position that since the operating system and BIOS were not published in a format that humans could read, and since there was no copyright notice in the software itself, it was not copyrightable. The lower court had agreed, but the appellate court disagreed with this position. This was probably one of the most important legal decisions impacting the evolution of the computer industry. It certainly was the most important decision impacting me.

Once the court's decision was reached in August 1983, the press treated the story as if we had lost the original case. That was not really what happened in technical terms; the lower court was instructed to rehear the case in light of the appellate court's instructions. While the court made clear that software, including the BIOS, could be copyrighted, it was not at all clear that Apple had copyrighted its system correctly. Apple had been very sloppy about such matters, and there was an excellent chance that Franklin would have won again in the lower court. Unfortunately, we would not have the funds to continue pursuing legal action.

— *either borrow a lot or borrow nothing*

I had been paying little attention to Franklin's cash flow. While I had learned a great deal about profit and loss at Mennen-Greatbatch and then at Digital, I had never really managed cash. It certainly had not been a consideration at Digital Equipment. Franklin was growing and was very profitable. We had a line of credit with Mellon Bank. They gave us an 80 percent advance on our receivables. So, we borrowed the cash we needed to grow our sales. As long as our sales kept growing, and they did, there was nothing to worry about—except that, the day after we lost in the appellate court, Mellon Bank called us and said they no longer wanted to extend the loan. They used the receivables to pay down the loan. They would not lend us anything more. Because most of our customers paid us between sixty and ninety days after shipping, we had no cash coming in. That's how I learned that you either have to borrow a lot—so a lender has to work with you to keep you solvent to get the loan paid off—or borrow nothing at all. Mellon did not care what happened to us; they would get paid, even if we went bankrupt as a result. It became apparent to me that we would run out of money. I was in shock. When our controller showed me how many days of cash we had, I threw up.

We took the only actions we could: laying people off and delaying payments to our vendors beyond the terms we had agreed. We prioritized the vendors that we most needed going forward. Vendors that were not strategic would not get paid at all. I started getting angry, even getting threatening phone calls at home. It was hell, and I was unprepared. At that time, it would have been a lot better to be experienced than smart.

We had a lot of components in our inventory: memory and CPU chips, floppy drives, hard disks. We started selling everything off at a discount, just to raise cash; we lost money on everything we sold. I was learning a valuable lesson about the difference between profit and loss and cash flow, a lesson that greatly helped me later when I was investing in early-stage companies.

— never count on a court ruling

It was so hard to know what to do after we lost that appeal. We had foolishly counted on a court ruling that we did not control. That was my first and last experience with the courts. We then initiated discussions with Apple, which had an incentive to settle because Franklin could still go back to the lower court. We reached an agreement in January 1984. By then, we had developed a version of the operating system and BIOS that did not infringe on the Apple copyright, but it was not 100 percent compatible. Combined with our fire sale of inventory to raise cash, this killed our sales. In the end, we paid Apple $2.5 million, a sum Franklin could not at all afford.

I WAS THE RIGHT GUY AT THE WRONG TIME

I thought of myself as a gifted strategist. Strategy involves just a few critical decisions. Now I was handling a thousand little choices with no relevant experience to help me make them. I was miserable, as if I had been blissfully driving a car at a hundred miles per hour only to suddenly realize with despair that not only was I running out of gas but also my brakes had failed.

When I joined Franklin, Jim Simons taught me a fundamental lesson: he said I should personally build a P&L (profit and loss) model of the company using a spreadsheet. He told me that if I made it myself, I would understand the company. He was so right. Building a model yourself is very different from getting one from the finance department. You learn how all the parts relate, and you can easily change values and see the results. That model of Franklin's P&L was the first spreadsheet I ever made myself. I used a Franklin Ace 1200 and VisiCalc. I only wish that Jim had also suggested I model the cash flow.

— sometimes being smart is not good enough

Jim invited me to dinner after our board meeting. I was looking forward to that, because I wanted to resign. I had hoped that Jim would ask me to quit, because I had a severance agreement and needed the money. Jim eventually said that he thought it would be best for the company and for me if I left. I agreed. We worked out the arrangements the next day. I think Jim liked me and respected me, but we both recognized that I was the wrong person to lead Franklin in its new circumstances.

On April 9, 1984, a bit less than thirteen months after joining Franklin, I resigned. This drew a lot of press. I was clearly the wrong person to navigate the company through the period that lay ahead. Barry Borden took over my responsibilities but was gone a month later. Joel Shusterman returned, at least for a little while.

After dinner with Jim, I went home and told Arianne that I no longer had a job. It was the first time in eighteen years that I was unemployed. I had no idea what the future would bring.

A GOOD DEAL FOR SOME

In June 1984, Franklin filed for Chapter 11 bankruptcy. The reason concerned the company's previous financing. The noteholders, who included Jim Simons, were able to force the company into bankruptcy, which wiped out all the shareholders, including the founders and employees. While Jim's shares were also wiped out, the fact that he held so much debt meant he ended up owning about 50 percent of the company.

Mort David joined the company in May as CEO. He put the company into bankruptcy and brought the number of employees down to 100 from 250. As part of the bankruptcy settlement, Franklin issued shares to its creditors. Because of the number of shareholders that resulted, Franklin ended up a publicly traded company without having to go through the process of an IPO. Those shares issued to vendors eventually became worth something. I remember calling a few vendors myself to make sure they knew, even though I had long since left Franklin. Unfortunately, the bankruptcy wiped out all shares owned by the funders, the team I brought to Franklin, and myself.

Franklin no longer had an opportunity in the Apple- and IBM-compatible PC market. David acquired a speech company, Proximity Technologies, and turned Franklin into a very successful company offering products like electronic spelling devices and electronic bibles. In 1990, the company's name was changed to Franklin Electronic Publishers. David left the company in 1999 after fourteen years as CEO. The only thing that remained of my time there was the logo.

Once the company entered bankruptcy, my severance agreement was terminated, which put me in a pretty uncomfortable financial situation. My million-dollar signing bonus paid in stock was worthless, but since Franklin had already paid the tax, at least I had a million-dollar long-term capital loss that took many years to use up.

WHAT IF THERE HAD BEEN NO LAWSUIT?

If Franklin had prevailed at the appellate court, the company would have gone public. Franklin was growing as fast as Compaq, and while we had some management issues, these were not unusual and would have been resolved. Likely, the market for the Apple II and compatibles would have rapidly expanded, providing serious competition to the IBM PC. Apple could also have decided to license their software and maybe could have been Microsoft. But this was not what Steve Jobs wanted. Steve was committed to a proprietary system of hardware and software, but the more open, horizontal IBM PC eventually won most of the market for personal computers only to see its market share decimated by clone manufacturers. Apple's much later development of the iPhone would demonstrate that a fully proprietary, vertically integrated system could once again be done successfully.

I sometimes have fun speculating about what it might have meant had Franklin seen success akin to Compaq. That could have meant eventually buying Digital Equipment Corporation, with me (in this alternate reality) as the CEO of the combined company. While unlikely, I nevertheless enjoy imagining that I could have ended up CEO of Digital.

I had hit a brick wall. I felt like a total failure; I had traded a big job at Digital for the title "former president of a bankrupt company making unsanctioned Apple clones." However, I now realize that my time at Franklin was a great learning experience. First of all, I gained some humility, a statement at which many of my readers will surely laugh. Laugh or not, I learned that just being smart is not enough; when things go wrong, experience matters. I also learned that I did not like being in charge of a company. I decided that I would never take on that role again, and I never did.

I had moved my family to the Philadelphia area to join Franklin, which had few technology companies. We would most likely have to move again, but we had only scarce savings. Arianne was not working, and we had three young children. I was scared but also relieved. I had never before been in a situation where the past had stopped while the future had not yet begun.

FINDING A NEW POSITION

At first, I took calls from headhunters about various positions. As Franklin got into more trouble, those calls became fewer. I did take a few interviews, but none of the positions appealed to me.

Then my close friend Steve Mayer, one of the founders of Atari, asked me if I could help him.

He was living in New York at the time and running an advanced development group for Time Warner called Warner Labs and living in a beautiful brownstone. Though the lab was developing several exciting projects, Warner was having difficulty and wanted to wind it down. Steve asked me to review their various products and technologies to see if there was anything worth spinning out. We found one project that we believed had promise, which became the foundation of the Emmy Award–winning Digital FX, which made workstations for special effects. I helped negotiate a wind-down agreement with Warner that provided Steve with some cash and the intellectual property that became Digital FX. Later, Steve Jobs would have Apple invest. Jobs had worked with Steve Mayer at Atari before he started Apple.

Jobs insisted that the company be located in Silicon Valley close to Apple. So Steve sold his town house in New York and bought a large home in the Los Altos Hills. So many companies, famously, have been started in a garage. We used to joke that Digital FX was no exception, although this was a three-car garage and part of a multimillion-dollar home in the Los Altos Hills, one of the most exclusive areas of Silicon Valley.

Steve paid me well, but doing this work, somewhat stupidly, just delayed my focus on getting a real job. Now I was really starting to panic. It had been three months since I'd left Franklin, and the company had just gone bankrupt, terminating my severance agreement, and I was no longer working for Steve. I had no money coming in and a lot of money going out. I had several potential but unappealing job opportunities. I held out, even though it was hard to resist taking just any position given my scary financial situation.

I had been on someone else's payroll for eighteen years, and I had only been out of work for four months. I knew I should have counted myself lucky, but those four months were truly terrifying. I had to find a way to provide security and stability for my family. That meant no start-ups. I wanted to work for a company with a strong balance sheet.

The company Convergent Technologies had attracted my attention. Allen Michaels, formerly of Digital Equipment (and later, Intel) founded the company in 1979, along with several people who had been at Xerox. The company sold workstations, as well as servers, based on Intel microprocessors. It seemed like a good fit with my background, especially given the work I had done at Digital. My friend Mort Ruderman introduced me to Michaels, who invited me to meet him at his Silicon Valley office and interview with various executives. I flew there on a Friday sometime in June 1984.

My interviews at Convergent did not go well. I talked with Michaels a bit. He had a big personality, and I think our conversation centered more on him telling me about himself rather than learning about me. I spent the most time with Ben Wegbreit, who ran one of the company's divisions. He would have been the hiring manager had I been offered a job. Ben was brilliant, very intense, and very aggressive. He had a strong academic background, which I obviously lacked. At the time, it seemed very clear to me that Ben was not really considering hiring me. Instead, he seemed intent on proving that I was not qualified to work at the company. It was excruciating and humiliating. It was so upsetting that I still remember it today. I honestly can't personally imagine treating anyone the way he treated me that day.

Years later, after I was well established at Intel, Ben reached out to me for some reason I no longer remember. I let the past go, and I have had contact with him a few times since. I never did tell him, unless he reads this book, just how much he hurt me. In a blog post I read recently, Steve Blank, who worked for Ben, describes his successful job interview at Convergent. The intensity of it sounds a lot like my interview. That makes me wonder if Ben was not trying to humiliate me—instead, I had just done too poorly in the interview to get hired. And thankfully so, or I might have ended up at Convergent and not Intel. In 1988, Convergent was sold to Unisys.

I felt so vulnerable after the Convergent interview. I was running out of money. After I left my meetings at Convergent, I called Arianne. I remember crying as I told her what had happened. I felt I had no more job prospects. I had a breakfast meeting planned the next day with Les Vadasz at Intel and had very low expectations for that meeting. As it turned out, Intel would be my home for the next fifteen years.

8. FAILING UPWARD

In the summer of 1984, I met Les Vadasz met for breakfast in Woodside, a small town in the hills above Silicon Valley, where many of the most influential people in the Valley lived. Woodside's restaurants and bars served as meeting places for venture capitalists and entrepreneurs. They still do.

Les was forty-eight years old, and I was thirty-nine. A rather slight man, and balding, he had a mustache and an unforgettable lovely smile. He has both to this day. Les welcomed me in his Hungarian-accented English. Like Andy Grove, Les was a Hungarian Jew who survived both the Holocaust and the Russian occupation of Hungary. Leaving during the short-lived Hungarian revolution of 1956, Les had made his way to Canada and enrolled at McGill University in Montreal. He moved to the United States in 1961. Around the same time, Andy Grove also left Hungry, landing in New York City. They did not know each other in Hungry and only met after Les joined Fairchild Semiconductor in 1964.

The story of Fairchild Semiconductor is essentially the founding story of Silicon Valley. In the early 1950s, William Shockley led the group at Bell Labs, the research arm at AT&T where the transistor was invented. He won a Nobel Prize, alongside the inventors John Bardeen and Walter Brattain, for that achievement. The transistor was arguably the most important electronic invention of the twentieth century, and its importance rivals that of the light bulb, invented in the century prior. So much of modern life involves the use of transistors. One of the greatest joys of my life was working with individual transistors and configuring them into devices that could actually do things. The chips used in our computers, phones, televisions, automobiles, and so much more are nothing but a collection of transistors. Billions of them.

Shockley left Bell Labs in 1956 to form his own company, Shockley Semiconductor Laboratory, as a division of Beckman Instruments, which financed Shockley's endeavor. Modern-day venture capital did not yet exist. Shockley moved from the East Coast of the United States to Mountain View, California, in what is now the heart of Silicon Valley, primarily because his elderly mother lived in the next town over, Palo Alto, the home of Stanford University. Amazingly, Shockley's mother may have been partly responsible for the creation of Silicon Valley.

Shockley, an excellent recruiter, assembled an outstanding team of scientists. The team included Bob Noyce, then twenty-nine, with a PhD from MIT in physics and mathematics, and Gordon Moore, twenty-eight, with a PhD in chemistry from Caltech.

Though a brilliant scientist, Shockley was a difficult and abusive man. When he started using lie detectors to check up on employees, Bob Noyce, Gordon Moore, and six others decided to leave and start a new company. Arthur Rock, the legendary venture capitalist who would later back both Intel and Apple, helped the team secure financing from Fairchild Camera. Fairchild owned the majority of the new company, Fairchild Semiconductor, and had a right to acquire the rest. They eventually did, making all the founders multimillionaires in the process. At this time, Arthur was working in New York for a boutique investment bank that focused on raising money for early-stage companies. Once Fairchild got going, he decided to move to the West Coast where he became a venture capitalist in the new world of high technology, which was just starting to develop in Silicon Valley.

It was at Fairchild Semiconductor, where Noyce, in 1959, invented the integrated circuit, with a lot of help from others. The integrated circuit was one of the most important inventions of the twentieth century. Before the integrated circuit, components such as transistors, diodes, and resistors had to be connected together externally by wires. The planar process that interconnected components without wires is actually a photographic process, which is why Fairchild Camera was not such a strange parent for Fairchild Semiconductor. Gordon Moore recruited Andy Grove to join Fairchild right after Andy graduated in 1963 with a PhD from the University of California. Les Vadasz also joined the next year, with a BS in electrical engineering from McGill. Later, in 1967, he became a strong advocate for developing a technology called "the silicon gate" that would later be used at Intel and become the dominant semiconductor technology.

At Fairchild, in 1965, Moore wrote a paper in which he described the doubling every eighteen months of the number of transistors held in the same size of semiconductor. He later revised this prediction to every two years, which represents a 40 percent compound increase in capabilities every year. Every five years, according to this prediction, the capabilities of semiconductor products increase by a factor of ten. No other technology in history has displayed this degree of improvement. While Moore made a prediction from an observation—it is not a law of nature—the idea became known as Moore's law. Moore's law fueled the technological revolution of the past half century.

Andy Grove quickly grew into a key and complementary member of the executive team at Fairchild Semiconductor. The company became very profitable. However, the more successful they were, the more interference they got from their parent company, Fairchild Camera. Furthermore, Fairchild Camera did not sufficiently reinvest their profits from the semiconductor business back into that business. As a consequence, Fairchild Semiconductor began losing its leadership in the very market the company had created.

Bob and Gordon decided to leave Fairchild to start Intel in 1968. They put in some of their own money from the payout they had received from selling their shares to Fairchild Camera, as well as additional investment from Arthur Rock. Andy Grove and Les Vadasz became the first two employees. Most people think Andy was one of Intel's founders, but he was actually just one of the first employees. Noyce had Badge #1, Moore had Badge #2, and Vadasz actually had Badge #3. Grove had Badge #4.

Michael Malone's book *The Intel Trinity* describes how Noyce, Moore, and Grove built Intel. I think it's an apt name, because creating a successful company requires three main ingredients. Noyce was the visionary. Moore was the strategist. Grove was the manager. These three capabilities proved critical to Intel's early success.

— vision, strategy, and execution are vital ingredients

Intel's initial target market was dynamic random-access memory (DRAM). At that time, computers used magnetic core memory. Memory components built using semiconductor technology, Intel believed, could be cheaper, faster, smaller, and more reliable. They were right.

Memory components, while requiring advance manufacturing technology, have little logic complexity, just storing a series of bits (ones and zeros). Once Intel introduced memory products, they would continue to increase the density (number of memory cells) and improve the reliability of these components, which also lowered the cost per bit of memory. Computer architecture was unimportant in this business. Semiconductor process technology was everything, and that was the expertise of most of the senior management. Les Vadasz was one exception, as he was an electrical engineer who understood a bit about how computers actually worked. Les even had some experience with programming.

While Intel intended to manufacture silicon memory components, they knew it would take some time for that business to develop. To augment revenue in the meantime, they undertook custom silicon design work. One such project was for a Japanese calculator company, Busicom, that Bob Noyce had found. Busicom wanted Intel to develop seven to fifteen custom chips. Ted Hoff, Intel's twelfth employee and a PhD from Stanford who Noyce had recruited, was assigned to work with Busicom. Hoff soon concluded that doing this project would be too difficult and expensive. Intel did not even have enough design engineers to do half the required work. But Hoff had a brilliant idea. He realized that it would be possible to create a general-purpose processor that could perform all of the needed calculator functions under software control. Busicom eventually agreed to this strategy. Federico Faggin, who had previously worked for Vadasz at Fairchild developing silicon gate technology, joined Intel, reporting to Vadasz to lead the development of the first microprocessor, the Intel 4004 in 1971.

While Grove is the Intel executive most often associated with the microprocessor, Noyce actually had the vision. In the early 1970s, he routinely told people that one day microprocessors would become computers. There was a big debate within Intel about the wisdom of marketing a general-purpose microprocessor. Some in management argued that many of their customers of memory components would see such a microprocessor as competition: their supplier, Intel, moving into their market. Computer companies built their central processors using very different forms of semiconductor technology. This was one of their core competencies. Now Noyce was proposing Intel take over this function in just a few chips that any firm could buy. Some within Intel argued microprocessors were simply a bad business concept, because customers would buy just one microprocessor compared to many memory chips per system. Others argued that having more processors would lead to a great demand for memory chips. Intel eventually decided to market the 4004 microprocessors. Noyce renegotiated the deal with Busicom, with Busicom obtaining a reduced price in exchange for Intel recovering the right to market the 4004. Ironically, Andy Grove was one of the senior managers who had opposed this. Andy was concerned that the 4004 would be a distraction that took focus away from the development of memory products. Andy was always in favor of focus. It was both his greatest strength and his greatest weakness.

As Intel became highly successful in the memory business, microprocessors represented just a fraction of the company's revenue. However, selling microprocessor development systems was a great business. The memory business became an easy target for competitors. The Japanese were particularly interested in this market, which played to their strengths as manufacturers.

In 1984, just a few months after I joined Intel, I was asked to drive Andy Grove back to the Portland Airport. Andy often flew up to Oregon, and managers rotated through the responsibility to either pick him up or drop him off. This gave Andy an opportunity to get to know these managers and hear their thoughts directly. Andy looked very pensive when I picked him up at the Intel office. He told me about the problems Intel was having with its memory business against the Japanese competition. He seemed frightened, and I'm sure he was. That would not be the last time I saw that look on Andy's face. He once said two things drove Silicon Valley: "greed and fear." We were clearly in the "fear" stage. Later, Andy would write a book on this theme titled *Only the Paranoid Survive.* Andy said as we drove that there was a chance that Intel would end up acquired by a Japanese company. I found the entire discussion pretty alarming.

Later that year, Gordon and Andy made one of the bravest strategic decisions ever made in the technology industry. They decided to exit Intel's DRAM memory business and focus on the nascent microprocessor business. Intel did keep some other types of memory components, such as EPROMs. This decision led to significant losses in the short term, including layoffs of large numbers of employees. Gordon and Andy realized that Intel did not have much of a choice; their memory business would be crushed either way. Nevertheless, many other business leaders would have waited to act until it was likely too late. Intel then entered its second act, moving from leader in memory products to leader in microprocessors. Until the PC took off, the military, automotive, and telecommunication industries were the main customers for Intel's microprocessors. None of the successful personal computers such as the Apple II, the Commodore PET, or RadioShack's TRS used Intel processors. Personal computers were not considered an important market for Intel.

JOINING INTEL

I knew little of Intel's backstory when I agreed to meet with Les Vadasz that morning in 1984. Among his other responsibilities, Les Vadasz managed Intel's complicated system of strategic planning, with people embedded in various organizations across Intel and reporting to Les, as well as to their organizations' managers. One of these people, Mike Richmond, who worked inside Intel's Systems Group, had read about me in an *EE Times* article discussing my resignation from Franklin Computer. Before joining Intel, Mike had worked at Data General,

Digital's primary competitor, so he knew the computer industry. I guess Mike felt my experience building computer systems was something Intel needed. It took several months for Mike to find me; this was not so easy in the days before the Internet. Eventually, his group's assistant tracked me down. Mike and I spoke by phone, and then he called Les. He suggested to Les that Intel should try to hire me. Les was interested.

Soon after, I spoke with Les by phone. We had an excellent discussion, but Intel at that time was primarily a memory-chip company. Though it supplied some microprocessors, including the 8088, the central processor for the IBM PC and its various clones, its main business at the time was dynamic random-access memory (DRAM). The IBM PC had been announced just two years earlier. Intel had less than a billion dollars in sales. The company seemed like a poor fit for me. I told Les that if I ever found myself in Silicon Valley, perhaps we could meet up.

Les would become my boss, partner, and friend. Of all the people I ever worked for, he undoubtedly had the most profound impact on my life. A brilliant, talented, and modest man, his ego seldom got in the way of doing the right thing. He was totally committed to Intel's success, and he was my greatest mentor. In my opinion, his role at Intel and in the development of the semiconductor industry overall is grossly underappreciated.

I can't remember much about our initial conversation, but when we met later, I knew that by the end of our meal, we had both decided that it made sense to explore the possibility of my joining Intel. Les told me that someone from the human resources department would contact me early the following week to arrange for my return for interviews. Leaving the restaurant, I found a pay phone and called Arianne to tell her that my meeting with Les had gone well and that I would soon return to the West Coast for interviews. Then I flew back to Philadelphia. The following Monday, as Les had promised, someone contacted me from the Intel HR organization assigned to handle my visit and potential employment.

I had three days of meetings scheduled, two in Oregon, where Intel had major facilities, and a final day in Santa Clara. Many of the people who interviewed me worked for Les, who was one of a few senior vice presidents. He had been instrumental in getting Intel into the microprocessor business. Elected vice president of component engineering in 1975, he then took over responsibility for the corporate strategic staff (CSS) in 1979, a fancy-sounding name for a collection of various activities that included computer-aided design, corporate planning, and new business development.

Moore and Grove agreed with Les that the company needed to bring in some high-level outside talent. Intel was very inbred. Les was told that he could hire one senior person every year. As it turned out, I would be his first and last strategic hire. As far as I know, I was the only successful senior person brought into the company from outside over my fifteen years at Intel. The 1998 Annual Report lists twenty-eight corporate officers. I was the last of these to join the company, fourteen years earlier.

It's not possible to remember all the people I met during my interviews at Intel. There were so many. In Oregon, I interviewed with Ed Slaughter, who reported to Les and was running an "incubator" for new businesses called the Intel Development Organization (IDO). I interviewed with Justin Rattner, who worked for Ed and was running an internal start-up building a supercomputer using multiple Intel CPUs. Justin was responsible for the design of the ill-fated 432 microprocessor. He eventually became CTO of Intel. I met with Mike Richmond, who had initially referred me to Les from that *EE Times* article, as well as Bill Lattin, who was running Intel's Systems Group. Although I was the person being interviewed, interacting with so many people in a short period of time provided me with a unique opportunity to learn about Intel, which proved important when I interviewed with Andy.

When I got to Santa Clara, Les and I spent a great deal of time talking. I asked him what my position might be should I be given an offer. Les said he had no idea. If I joined, my first task would be to figure that out. Many people might have been uncomfortable with such a statement, but for me, it was perfect. During our discussion, he explained the method Intel used to do strategic planning. Every year, a document called the strategic long-range plan (SLRP) was the result of a multiday planning session attended by senior management and a few others. From 1988 on, I would participate in the SLRP sessions, which allowed me to see and influence Intel's strategic thinking up close. I asked Les if I could read the current plan. He agreed but said that I had to read it in a conference room and could not take notes or remove it.

— a great strategy executed badly
looks like a bad strategy

On the first page was written, "Change is our ally, we should embrace it," or something to that effect. I couldn't believe it. I was thrilled. That one phrase had a great deal to do with my decision to accept Intel's subsequent offer. That I was running out of money was also good motivation. I wanted to become an agent of change. My experience working at Digital Equipment, however, also made me realize the absolutely vital importance of execution. I had learned that a great strategy executed poorly would end up looking like a bad strategy. From my many meetings across Intel, I saw that Intel was an "execution machine." I also thought the company could benefit from my strategic insight. By influencing Intel's evolution, I could impact the computer industry. First, though, I had to get myself hired. Les reported to both Gordon Moore and Andy Grove. He had explained to me that Andy's opinion would determine whether or not Intel made me an offer.

After Les, I met with Gordon Moore, Intel's CEO and co-founder. Gordon is soft-spoken and brilliant. Knowing and working with Gordon was one of the great honors of my life. Then, finally, I had my most important meeting, which was with Andy Grove, the president and COO of Intel. I had briefly met Andy in 1982 at the PC Forum when I was working at Digital Equipment Corporation. He then looked more like a movie star than an executive, with a Tom Selleck mustache and a large gold chain around his neck. Little did I know at that time the role he would play in my life.

Since I knew the meeting with Andy would determine whether or not I got an offer, I thought about what I could do to get his support. Andy had written a very successful business book, *High Output Management*, first published in 1983. Prior to my meeting with him, I read that book from cover to cover several times. I got a notebook and filled several pages with notes made while reading his book. I added several pages of questions I intended to ask Andy. Finally, I wrote a few pages of suggestions for Intel that I would offer to Andy should I get the chance.

Though my questions and suggestions were very real, the notebook was just a device I decided to use to get Andy comfortable with me. I never used notebooks, and I never took notes. I had an extraordinary memory back then, and I did not need to write things down. However, I knew that at a company like Intel, which was driven by engineers, notebooks and notes would be critical to demonstrating my seriousness. After I joined Intel, I would sometimes take a notebook into meetings and write down whatever I thought was significant so that others would know that I was paying attention. I never looked at those notes again.

When Andy started talking with me, I carefully wrote down things he said in my notebook. Then, I asked if I could ask him some questions. He agreed. I went to the section in my notebook where I had carefully written out questions and made sure he could see that I had done that in advance. After he answered my questions, I mentioned that since I had interviewed so many people, I had perhaps a few ideas that might be useful. He was interested in hearing them. I guess he liked all of that, because at the end of the meeting, while I was still in his office, he called Les and recommended that Les hire me. It was probably the first and last time he responded to me so positively.

During the interview, I thought Andy had said that Intel needed people like me who could help the company change and evolve. Over the years, I realized he must have said "make the company change," not "help the company change," because it was always a battle. Many years later, I attended a meeting with Andy and the CEO of Sony, Nobuyuki Idei. At one point, Idei asked me to describe my role at Intel. I said, "My job is to make Intel do things it does not want to do." He looked surprised. Later, Andy said, "I can't believe you said that." I replied, "It's true." And it was true. I just wish that I had been able to get Intel to do more things it did not want to do. Andy and I would have a complex and often challenging relationship. Either Andy was not that open to new ideas, or I was not that effective in presenting them. Both of those statements held true, I suspect. Andy would later call me Intel's Wild

Duck and said he always needed an Avram Miller on his team. Alas, this Wild Duck would eventually fly off.

I wrapped up my discussions with Les, and he said that he would like to offer me a position. We discussed compensation. He said that Intel would match the salary and bonus I had at Franklin and provide stock options. I accepted his offer. Les said that I could choose whether I wanted to work out of Intel's facilities in Silicon Valley, Oregon, or Arizona. I told Les I wanted to bring my family out to explore both Silicon Valley and Portland, Oregon. I felt really bad about moving my family from the Boston area to Philadelphia when I went to work at Franklin. Having lost my job there, we had no choice but to move again. And so Arianne, our three children, and I visited Portland and Palo Alto, the two cities we considered. We had a family dinner, and I left the decision up to the rest of my family. They decided we should live in Portland. We could afford a large and beautiful home there, which would not have been possible in the already very expensive Palo Alto.

It still amazes me that Intel offered me a job, especially at such a senior level. I had only worked in the computer industry for five years. My work at Digital had been anything but a success, and by the time I interviewed at Intel, Franklin Computer had gone bankrupt. I had no college degree at a company founded and heavily staffed by PhDs. I had not attended university. I'd barely made it through high school.

Recently, I asked Les why he hired me. He said it was because I was smart and creative, and he was looking for someone with whom he could interact. Happily, that turned out to be me.

INTEL: A MINICOMPUTER COMPANY?

My start date was August 6, 1984. I came out to California without my family and spent a few weeks at Intel's main site, in Santa Clara on Bowers Avenue. Intel denied having an actual headquarters, but this Bowers address sure came close to being one. Not only did Gordon Moore and Andy Grove have offices on-site, but even Bob Noyce, then the chairman of the board, still had an office there. Starting there gave me an opportunity to meet a lot of the company's key executives.

Les and I decided that since I would be based in Intel's Hillsboro, Oregon, facility, I should attend staff meetings of the Systems Group that was headquartered there and see if I could help them. I discovered, much to my surprise, that this group was developing personal computers, which they planned to sell as a "white label" product to companies wanting to offer their own branded PCs without designing or manufacturing the machines. I carefully reviewed the group's engineering plans, especially the cost structure. My experience at Digital and then at Franklin made this an area I knew well. I quickly concluded this plan was a serious mistake. Not only would the Systems Group be creating friction with Intel's microprocessor customers, but also they were unlikely to succeed in the market. Even if they did, it would not be very profitable. Through my industry contacts, I got pricing from other PC manufacturers that were willing to build private-label systems. I demonstrated that the prices for those systems were less than the cost of Intel building an equivalent product. As it made no sense to be in this business, the project was terminated. Later, the Systems Group would try this same thing again. That time, they actually succeeded in getting a few major design wins, including AT&T. The arrangement worked out poorly for both companies.

By the mid-1970s, Intel realized the role that the microprocessor would play in the future of the computer industry. It was not so obvious at that time. While mainframe and minicomputers used semiconductor technology, they did so in a very different way. Gordon and Andy advocated for the development of a "mainframe on a chip," a microprocessor that could serve as the heart of computers. However, at that time, they did not understand the changes that would eventually transform that industry and Intel's place within it, driven by the very same technology Intel was developing. They did not anticipate the size of the PC market at all, then less than a decade away. They believed that the computer industry would continue to be dominated by proprietary and expensive mainframes and minicomputers, such as those offered by IBM and Digital Equipment. Though the first personal computer, really a hobbyist's machine, the Altair 8800, was released in 1975 and used an Intel 8080 as its microprocessor, Intel did not see personal computers becoming an important market. Atari had inquired about using the 8080 when they started to design their home computers. But Intel showed little interest in working with Atari on a consumer product, so Atari selected the 6502 microprocessors from MOS Technology. This led Apple to select the 6502 for the Apple II as well, since Jobs, who had worked at Atari, was heavily influenced by that company's decision. The Apple II, the TRS-80 from RadioShack, and the Commodore PET, as well at the Atari Video Computer System, were all released in 1977. None of these used an Intel microprocessor.

Back in 1975, Intel had embarked on a significant project to develop that mainframe on a chip. At first, Intel called it the 8800, and later renamed it the 432. Justin Rattner, a young computer scientist who had joined Intel in 1973, proposed that rather than developing a 16 bit to follow the 8-bit 8085, Intel would develop a new advanced architecture that would incorporate the best academic concepts in computer design. Gordon was tasked with figuring out which microprocessor architecture would be used for future processors. He formed an internal committee with three members and brought in Carver Mead from Caltech to advise him. The committee failed to present a unified recommendation. Realizing that there was a lot of risk related to the development of the 432, it was decided to develop in parallel a 16-bit follow-on to the 8085 called the 8086. That product was released in 1978. The 432 would not be released until 1981.

Development of the 432 started at Intel's Santa Clara site under the leadership of Bill Lattin, recruited from Motorola to lead this project. The group later moved up to Oregon, where Intel had already established a manufacturing group and where Lattin had grown up. In the meantime, Intel began the development of 8086 in Santa Clara. Later, a critical decision to develop an 8 bit version called the 8088 was made. This was the chip that IBM would choose for its PC because it used the same I/O bus as the Datamaster, a word processor that IBM was developing. Intel was facing a great deal of competition. Had IBM not chosen the 8088 for its PC, it is likely that the company would have never been a leader in microprocessors. Likely, the company would not even exist now.

One of the better marketing minds around, Bill Davidow comanaged the microprocessor business along with Les Vadasz. It was under Davidow's leadership that Intel had begun to offer development systems, which simplified the design process for engineers wanting to use Intel's processors. Perhaps Davidow's greatest contribution to Intel was to get the company to understand that the product was not the device. Rather, the product was everything the company offered to make the device successful, including development systems, documentation, technical support, and reference designs. All of this played a very important role in Intel's success in the microprocessor business. Intel beat Motorola despite having an inferior device (microprocessor), because all the other capabilities won customers over. The focus on design systems also led Intel to develop an expertise in software. Intel was way ahead of Microsoft with respect to system software at that time. Had it wanted to, Intel could have been both the provider of the processor and its operating system. But it was concerned that if it sold the operating system, that would result in greater completion in the development system market. This was before the PC changed everything.

Lew Eggebrecht, the architect of the IBM PC, told me personally that Intel's development systems, including its compilers and debugging systems, were the most critical factor in his decision to select the Intel 8088 for the IBM PC. In 1979, Intel was losing the microprocessor market badly to Motorola. Davidow complained to Grove, saying he could not sell design systems if there were no design wins. Grove said to Davidow, "Then you go fix it." With Grove's support, Davidow organized the company into something called Operation Crush, which targeted two thousand design wins. Many believe that Operation Crush resulted in the design win of all design wins, the IBM PC.

I do not believe Crush had anything to do with IBM's choice, although others do. The 8088 was chosen for at least two other good reasons. First, the 8085 was being used in the IBM Datamaster, and the IBM PC development group wanted to use much of the motherboard design from that product. The second reason was Intel's superior design tools.

Besides development systems, Intel also offered memory boards. I remember buying them when I was with Mennen-Greatbatch in the 1970s. Eventually, Intel decided to consolidate its computer systems activities into one group, the Systems Group, headquartered in Oregon and led by Bill Lattin, as I mentioned. That group developed multibus, which became an industry standard bus architecture. The Systems

Group offered various products built on this architecture, including single-board computers, most of which went into embedded systems in addition to development systems.

It is astonishing to realize that the computer architecture forming the foundation of the personal computer market was not really considered strategic. Back then, microprocessor sales represented less than 10 percent of Intel's revenue until the IBM PC took off in 1981. The primary markets were automotive, military, and communications. Total global sales of all microprocessors for all companies surpassed only $500 million by 1984. The 432, meant to be Intel's primary processor, was, it turned out, too complicated for both the market and existing semiconductor technology.

Intel's founders, including Andy, really did not understand computer architecture, or even digital logic. They were semiconductor process people: chemists and physicists. Intel's primary business until 1984, memory chips, have a much simpler architecture than a microprocessor and require no software development. This meant Intel's executive management lacked the experience and skills needed to evaluate the feasibility of a project with the complexity of the 432. The only exception to this was Dave House, who came to Intel with a background in computer systems and who was one of the people who spoke out against the development of the 432. The failure of the 432, coupled with declining margins in the memory business, should have led to an unhappy ending for Intel had the IBM PC not saved the company.

Before the PC took off, Andy Grove came to believe that 50 percent of Intel's revenue should come from selling systems. The computer industry was still vertical, and the system companies like IBM and Digital Equipment were very profitable. In 1980, following Grove's lead, Intel decided to create a group to develop a minicomputer system to compete directly with the computer systems companies. Les Vadasz was tasked with figuring out how to get this done. He asked Sudhir Bhagwan, who was working in the 432 group, along with one of his direct reports, Ed Slaughter, to help develop a project proposal. In 1981, Bhagwan and Slaughter presented their plan to Noyce, Moore, and Grove. They defined a family of multiprocessing, distributed, and fault-tolerant computers that would have significant competitive advantages over machines from companies like Digital Equipment and

Tandem. But the investment of $100 million required for their development was way beyond the range that Intel could afford, given its just $350 million in annual revenue at the time. Grove asked the team to find a partner that could put up 90 percent of the cost for 50 percent ownership.

Intel was in exploratory discussions with companies such as Kodak when Siemens, the large German industrial conglomerate, showed significant interest. That company's industrial group was an early design win for the 432. Though they realized that Intel's 432 would not be successful, they still needed to develop a new, highly reliable computer for factory automation and nuclear energy plants. Siemens also had a fairly unsuccessful commercial computer business, and they had a lot of money.

An agreement was formalized between the two companies in 1982. The project was called Gemini, signifying the two companies involved. Each company would contribute employees to work on the project. Siemens underwrote the computer development, and Intel financed all the chip development. Oregon would be the main development location. Siemens appointed Udo Offer, one of their employees, as general manager. Intel recruited Larry Wade from Digital Equipment as assistant GM. Then in 1983, two senior Intel executives, Casey Powell and Scott Gibson, left with fourteen other Intel employees, including Wade, to start a competitor to Gemini, Sequent Computer Systems. The Sequent people wanted to take a different technical approach from Gemini to pursue the same market opportunity, a simpler approach using an existing microprocessor from National Semiconductor. Later, they would switch to Intel's 386 processor and use the Unix operating system. A commercial success for a while, Sequent had a strong partnership with Oracle, the leading database company, and was eventually acquired by IBM in 1999 for $850 million.

With the departure of the Sequent founding team, Bhagwan took over as assistant general manager, and Randy Young led Intel's chip development. The heart of the Gemini system was the i960, architected by Glen Myers. It was the first microprocessor that could execute multiple instructions at once. It was a 33-bit computer, with the extra bit used as a parity check. Using the parity check, multiple processors could run simultaneously, and the computer could not only detect but also fix any failure. The computer architecture was built to take advantage of then new object-oriented programming language, Ada, which the US Department of Defense had commissioned to standardize its embedded systems.

The resulting architecture was elegant but incredibly overengineered. It made the 432 look simple, which was no easy feat. The project required all-new hardware, including four highly complex chips, a new operating system, and compilers for several programming languages.

I spent a lot of time understanding the Gemini project, which was dominated by engineering people. No businesspeople were involved at all, just three engineering groups working together to develop the hardware, software, and microprocessor. I reluctantly concluded that I needed to take responsibility for the project's business aspects, at least from Intel's perspective. When I joined Intel, I had the ambition to expand Intel's business into the consumer market. Though this was the opposite, I felt it was my duty because I knew more about the computer industry than anyone else at my level in the company. My reluctance was justified.

— revolutionary products rarely succeed

Les and I discussed Gemini's needs, and I laid out my case for taking charge. I was about to get involved once again with an overly ambitious project like the Professional that I had led at Digital Equipment. Sadly, this one was a lot less fun.

I moved my office to Jones Farm in Hillsboro, which housed the Gemini project. I hired an assistant and began to build a team. I would be located there for the next three years, and it was not a very rewarding period of my career. I thought that I had perhaps once again made a significant mistake by accepting this job at Intel and moving my family to Portland. Working on high-end computer systems sold to businesses and institutions was precisely the opposite of the sort of project that most interested me, and I was rather isolated from mainstream

Intel. Since I was also responsible for component sales of the i960, I did have some contact with Intel's component group.

These were challenging years for Intel, which was reflected in its stock price. The price of Intel stock when I began my tenure fell by 40 percent through 1987. The stock then nearly tripled in 1987 before dropping 50 percent in the Black Monday crash of October 1987, plunging alongside my dreams of sending my children to university. Fortunately, by the middle of 1988, when my oldest son, Adin, turned eighteen, the shares had recovered to be worth more than 50 percent more than the strike price of the options I had been given when I joined the company. I could afford to send him to university. By the time I left Intel in 1999, Intel stock had gone up more than fiftyfold. I held on to the bulk of the shares until just before the tech crash in 2000, so I experienced another major increase in the value of my Intel shares. The fifteen years I worked at Intel left me financially independent, for which I am very grateful.

We identified two significant markets for Gemini systems: transaction processing and industrial control. By the time I joined, Intel had abandoned the idea that the P7 (i960) would be Intel's mainstream processor, given the runaway success of the IBM PC, which utilized the x86. We therefore focused our component sales on markets that needed high reliability, such as the military, avionics, and high-end workstations. We even tried to get Steve Jobs to use the P7 for NeXT. We successfully got Boeing to use the P7 in its first "fly-by-wire" airplane, the 777, which would take flight in 1994.

BiiN

The original development agreement between Intel and Siemens left both companies free to compete with each other using the same computer technology. This structure would have created many problems. I knew that without a change in structure, our partner would soon become our competitor. So I proposed to Intel and Siemens that we create a joint-venture company to avoid this result. I negotiated with both parties for the better part of a year, requiring me to take monthly trips to Germany to meet with my counterpart at Siemens. The agreement was significant for both companies, and we got the required approval from both companies' boards. During the ongoing negotiations, I also worked on the branding for the new company, which we decided would be called BiiN. The new company was announced in July 1988. By that time, I had already left the project and moved to work in Santa Clara doing corporate business development, which ultimately led to the creation of Intel Capital.

Both Siemens and Intel wanted to bring in a world-class, well-recognized CEO to lead BiiN. They hired an executive search company and started looking. I don't think I was ever considered for the job, nor would I have taken it, if asked. Thankfully, I was done with big computer systems. They recruited Joe Kroger, who had been the number-two person at Unisys, a company resulting from the merger of Sperry Rand and Burroughs—two old and tired computer companies. Joe was a big-company guy, and BiiN was anything but a big company. Joe probably did his best, but he was not dealt a good hand. He was fundamentally a sales guy, but BiiN had nothing yet ready to sell. Internal political issues at Siemens further complicated BiiN's position; a competing effort within Siemens emerged that involved Sequent Computer Systems. Intel became totally focused on the PC market and lost interest in BiiN. In 1989, both companies pulled the plug. BiiN was no more. Six years of work by hundreds of engineers and others were flushed.

When the company was closed down just a year after launch, many within Intel would joke that BiiN stood for "Billions Invested in Nothing." This was not exactly true. While Intel could not recoup its significant investment directly, given Gemini's commercial failure, Gemini had recruited many talented computer scientists, programmers, and engineers to Intel. After BiiN closed, Intel would create the Intel Architecture Labs (IAL) to provide a home for all of this talent, led by Ron Whittier, a key Intel executive. Very skilled people like Kevin Kahn, Fred Pollack, and Steven McGeady, who had joined Intel to work on Gemini, would play very important roles in Intel. IAL was essential to Intel's further evolution, and I would eventually work closely with this organization later in the development of residential broadband.

For Les, BiiN's failure must have been incredibly painful. Gemini had been his baby, started at a time when all of Intel's key management believed that the future of Intel was to sell computer systems as well as semiconductors. But the project's complexity had led to significant delays, and its target market was no longer viable. Intel nevertheless continued to pursue the systems business for quite some time, particularly in the areas of networking and communications.

Though Les and Andy appreciated my work putting together BiiN, I considered it three years of wasted effort on my part. But my efforts had been successful, and I had demonstrated my ability to get things done. However, I should have seen the impossibility of launching a new proprietary computer architecture when the world of computing was being radically transformed by the PC. Like so many, though, including Intel's top management, I was slow to grasp the significance of what was transpiring: the restructuring of the computer industry.

But as it happened, I now found myself in the right place at the right time. I moved to Silicon Valley and began the most significant period of my career, as vice president of business development at one of the most important technology companies in the world.

9. DEALMAKER

During my period working on Gemini/BiiN (1984–1988), I had let Andy and Les know that I was not happy with my role at the company and even started to explore leaving Intel. Andy said two things to me that, as it turned out, were very important. First, he said that I should stay at Intel because the company accepted me for who I was. While it was a strange comment, Andy was right. Even though I was very different from the other Intel executives, I was always treated with respect. That was, I suspect, because we all shared many of the same values: a focus on integrity, commitment, honesty, and results. Second, he said that if I ever needed something personally, I should just ask.

CALIFORNIA, HERE I COME

As it happened, a few years later, my wife, Arianne, was accepted into Stanford's PhD art history program. We needed to relocate to California. Intel took care of all the expenses, even though the move was at my request. Arianne and I found a beautiful home in Palo Alto, which, by then, we could happily afford. My eldest son, Adin, went off to Brandeis University, so there were now only four of us. My office was located close to Gordon Moore and Andy Grove, fortunately, because this gave me a great deal of access.

Back in 1986, Les had taken over the Systems Group from Bill Lattin when Bill retired, and Les spent much of his time in Oregon where that group was headquartered. My new job was doing business development for that group, still reporting to Les. I was responsible for activities like acquisitions and minority investments in support of the various business units comprising the group. In some cases, this structure worked well; in other cases, not so well. The business units were very focused on the short term. They rarely came to me with deals, so I had to find opportunities that fit in with whatever they were doing. I, then, had to convince them to sponsor the deal. They tended to have little interest unless there was a short-term payoff. I was more interested in having a long-term impact, so this was disappointing. One group where I was successful in assisting was PCEO (the PC Enhancement Operation), an internal start-up that Les had originally sponsored.

My timing was perfect, it turns out. Intel was about to undergo a profound transformation, from being a manufacturer of semiconductor components to one of the leaders in the computer industry. The 1987 annual report title was "Intel the Architect of the Microprocessor Revolution." In the process, Intel would become one of the most valuable and influential companies in the world, and I would finally be able to play a meaningful role in making that happen.

STRATEGIC LONG-RANGE PLANNING

In 1987, the year before I moved to California and just twenty years after Intel's founding, Andy Grove was named CEO of Intel. He replaced Gordon Moore, who became the chairman of the board. Andy held the CEO position for eleven years and then became the board chairman until 2005. Grove, a Jew who had survived both the Holocaust and the Russian occupation of Hungry, and who'd come to the USA at the age of twenty-one not speaking English, would be hailed as one of the greatest CEOs of all time, if not the greatest. Much has been written about this remarkable man, most of which I will not repeat here.

In the '70s, Les Vadasz was tasked with creating Intel's strategic planning process. One element of the process was called SLRP (strategic long-range plan). SLRP was initially bottoms up. As the company grew, SLRP became unwieldy, with more and more business units each presenting their own vision and plans. Intel had also evolved from being a semiconductor manufacturer of memory components into the world leader in microprocessors and, as Andy would describe it, "the foundation of the New Computer Industry." In 1988, Andy changed SLRP to be top down. The result was that Andy now became Intel's strategist. A consequence of this was that Intel's focus became narrower and deeper.

The SLRP meeting typically lasted about three days and would take place in April. It had about fifty attendees, including Intel's senior leadership and individuals of their choosing. Attendees were broken down into subgroups to work on various topics Andy presented. Subgroups would then review the results of their deliberations with the larger group, followed by a discussion with the full group.

It was meant to be a three-year look into the future, with the resulting plan that covered the same time frame. I used to joke that the SLRP discussions would end up dealing with the last year, the current year, and the next year. Even if it is a three-year view of the future, three years is not long enough to identify the potential discontinuity that could either provide enormous opportunity or possibly disrupt the current business. I am sure that those who were responsible for the development of Intel's semiconductor technology had a much longer-term perspective, but this was not a subject for SLRP.

Andy began with a two-hour presentation reviewing past SLRP discussions, then outlining his views of the future and raising various issues and opportunities as he saw them. How I wish I had copies of those presentations! They reflected his superb analysis of the industry.

Andy, like the rest of us, had come to slowly understand that the personal computer launched by IBM in 1981 had effectively turned the computer industry on its side. This change presented both Microsoft and Intel with the opportunity to dominate the computer industry. Something Andy clearly recognized.

However, over the years, Andy got stuck, and Intel got stuck along with him. Though Intel was enormously successful during the 1990s, it did not use either this time or its wealth to adequately prepare for the future that would result from the Internet and, later, from mobile computing. Intel even had difficulties initially grasping the importance of notebook computers and the growing home computer market, in my opinion. The company became fixated on the performance of its microprocessor. Moore's law would be the dominant force driving the company. Intel interfered that as meaning ever-faster microprocessors. Increased performance required more power, which resulted in shorter battery life and made Intel less competitive in the portable market. Even though Intel no longer had to enable second-source suppliers of its products, companies like AMD, and later Cyrix, and many others, offered compatible microprocessors. Intel therefore still had competition, especially on the lower end of processor performance. Intel's superior semiconductor technology meant it could build higher-performing processors than other companies. Processor performance was a key aspect of Intel's competitive strategy, but performance only matters to customers as long as there are desirable applications that can take advantage of it.

I find it ironic that Andy, who so clearly documented the importance of "strategic inflection points" for subsequent generations of business leaders, would end up missing so many. Andy used to say that two things drove Silicon Valley: paranoia and greed. Over time, Andy moved from paranoia to greed, from risk-taker to risk avoider. Eventually, Andy recognized that the Internet was such an inflection point. I think that was one of the reasons he turned over the reins of the company to Barrett. While he recognized it, he did not know what to do about it. Barrett had been very vocal about Intel's inability to create new business. Living in Arizona, he likened the PC microprocessor business to a creosote bush that would prevent other plants from growing. But while Barrett understood the problem, he was not the right person to come up with the solution. Nor were any of his successors.

VEIN OF GOLD

When the IBM personal computer was launched in 1981, very few people foresaw that the control of the computer industry would move from system manufacturers to two key component companies, Intel and Microsoft, resulting in great transfers of power and wealth. IBM certainly did not understand what it meant to be dealing with the likes of Andy Grove and Bill Gates.

Back then, Intel believed that the profit in the computer industry would continue to be made by computer system companies like IBM and Digital Equipment. Microsoft thought that Unix could be the dominant operating system and was working on a version of it, Xenix.

But as the PC market grew, Grove and Gates decisively abandoned their existing plans and fully embraced the PC opportunity. Over time, both companies came to realize that their customers were not the PC manufacturers such as IBM, Compaq, Dell, and Gateway but rather the individuals and enterprises that purchased those companies' products. Intel's marketing moved from focusing on the engineers that would design personal computers to the individuals and enterprises that would use them. This resulted in marketing programs such as Intel Inside, which created a very strong ingredient brand. Microsoft also developed a robust branding effort.

Fueled by growing profits, Intel was able to invest in developing much of the critical technologies required to move the center of the computer industry to the PC. Intel and Microsoft were then able to extract most of the profit in the computer industry.

Exploiting the PC opportunity was brilliantly executed, and Andy was able to get the company totally focused. He had found a vein of gold and was committed to extracting every ounce. But he did not realize that after you extract all the gold, you find yourself in a very big hole. Dominating the microprocessor used in the personal computer became known as Job 1. All other businesses were called Job 2, but after a while, Job 2 came to mean doing things that would support Job 1. While Andy wanted to see Intel develop new businesses, he offered his support for that objective inconsistently.

Intel's performance under Andy's tenure was astounding. This kind of success blinded Andy and the Intel board. With the exception of the period representing the collapse of the Internet bubble, sales and profitability continued to grow significantly for many years.

A N D Y   G R O V E : C E O

	Sales	Profit	Capitalization
1987	$ 1.9B	$ 0.24B	$ 4.3B
1998	$ 26.3B	$ 6.10B	$ 197.0B
Growth	13x	24x	45x
I N T E L P E R F O R M A N C E			

Andy Grove's amazing performance as CEO

I have grown to believe that what makes the long-term difference is not only what you do when things go badly but also what you do when things go well. It is a great challenge to develop new businesses in the shadow of any business as successful as Intel's microprocessor business. Intel, like so many other companies, was failing success.

Andy was an amazing manager, and he taught others in the company how to manage effectively. Much has been written about Andy's management capabilities and his management style, from Stanford Graduate School of Business Professor Robert Burgelman's *Strategy Is Destiny* (2002) to *Measure What Matters* (2018) by famed venture capitalist John Doerr, which details one of Andy's most effective management tools, objectives, and key results. Without a doubt, Andy was one of the best managers of any major technology company anywhere in the world, perhaps even the very best.

Andy deserves recognition for much of Intel's success, but along with that must come responsibility for its failures. While he was razor focused on exploiting the opportunity the PC presented, he did not lay down the foundation for continued success, including picking the right successor. Most of his fans and many of my former colleagues and friends at Intel will disagree with me. My example of a great CEO would be Steve Jobs, even though I think Andy was unquestionably a much better human being.

What Andy lacked was vision. For me, a great CEO has to have a combination of vision, leadership, and management ability. Yes, Andy would come to recognize the importance of the Internet and the changes that would result in the nature of the computer industry, but it was too late. He failed to develop an organization that was able to embrace the future but instead clung to the past. Nowhere would that become more evident than when Paul Otellini, then CEO, would turn down Apple's request that Intel manufacture the processor for the iPhone. Of course, Paul could not have foreseen the impact of that decision at the time. But he should have realized that things would not stay the same and that there would be disruptions. Apple was already very successful with the iPod.

While this book is more critical of Andy than many other books written about Intel, I should be clear that while he and I had a difficult and complicated relationship, I had the utmost respect for him professionally and personally. I think Andy, if he were still alive, would have agreed with much of my criticism. A year before he died, suffering terribly from Parkinson's disease, he apologized to me for not listening more to what I said. This meant a great deal to me. But I have to take responsibility for not getting him to hear me. In writing this book, I came to realize part of the problem. Andy was comfortable with the concrete. He wanted facts. But in thinking about the future, one has to use intuition and imagination. Andy wanted to know what I knew, not what I believed. But how could I know the future? I also believed you could influence the future. While you could not control it, you could bend it. One of our biggest disagreements came about because of our divergent views on the role of telecom companies in creating residential broadband.

— the future is too important to be left to chance

EXPANDING INTEL'S BUSINESS

Initially, in my new role, I had difficulties being effective. While I shared Intel's values, I struggled culturally; how I looked, dressed, and acted were just so different from the other Intel people that I was often not taken seriously. This reminded me of my difficulties as a young child when I returned to school after my year at the convalescent home at the age of eight.

At one point, I asked the human resources department if they could find someone who could help me adjust my behavior so I could be more effective. Intel was working with a psychologist who specialized in understanding corporate culture. After a few sessions with me, he said that I was too enthusiastic about my proposals and did not appear to be balanced by not presenting the downside. I was, in other words, too positive. He also thought I was relying too much on my intuition, not enough on objective data. This was also true, but I was strong believer in instinct. Instinct is when you know something but don't know how you know it. Not knowing how you know something does not make it not true. But in the highly analytical engineering culture of Intel, it certainly made it suspicious.

After working with the psychologist, I learned to be much more concrete and balanced in my presentations. I gave as much time to the downside as the upside, and I kept my enthusiasm at bay. His advice was great. Over time, as I had more successes, my views were given more credibility inside Intel, and I could revert back a bit.

While doing the research for this book, I interviewed many who worked either directly or closely with Andy. All held him in high regard. Often, they would say that while Andy could be very opinionated, he could be persuaded to change his mind when confronted with facts and data. This made me realize that one of my issues with him was that I could not present facts and data about how I thought things would unfold in the future. I was using my intuition.

It turns out that I did not have much to add to Intel's existing business, because my background was in computers and not in semiconductors. I am slightly embarrassed to admit that I never once set foot in a semiconductor factory during my time at Intel. Naturally, I looked for opportunities that fit my skill set; in retrospect, I realize this limited my impact. In Intel's results-oriented culture, writing reports and making presentations would have little impact. I needed to take action and make things happen. I would need to become an "activist strategist." But how?

It is very important for companies to expand their business. Successful companies usually begin with one product. Over time, they add additional products, which they sell to existing customers, and they acquire new customers/markets for existing products, including geographical expansion.

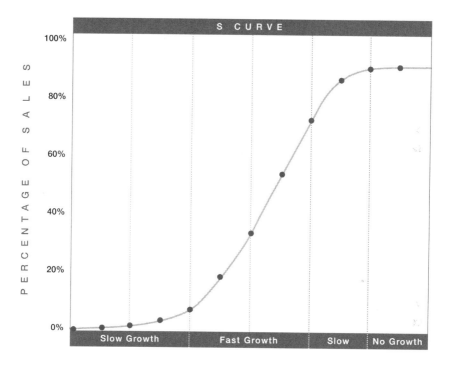

Products normally go through what is called an S curve. In the beginning, a product may not have enough features/capability to attract many customers. As capabilities are added or, sometimes, as prices are lowered, there are more and more customers. After a while, all the potential customers have bought the product and the market turns into a replacement business. It becomes harder and harder for a company to add capabilities that are so compelling that customers will want to upgrade. Sales level off and can even decline. Generally, the time for a company to add new business is when the current business is experiencing rapid growth. That is difficult because the business that is growing rapidly consumes most of a company's resources and attracts the best talent.

VERTICAL vs. HORIZONTAL INTEGRATION

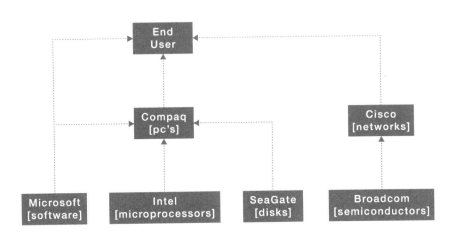

Graph illustrating vertical integration

Companies have a number of options for expansion. They can do a vertical integration; networks would develop to be formidable business. Intel's strength with the semiconductor and its software skills could play an important role in developing a leadership position in the nascent computer-networking business, I thought. Microsoft was struggling with networking technologies and was not entrenched. Furthermore, the development of computer networks would also help accelerate the adoption of personal computers to the benefit of Intel's main microprocessor business. At first, it seemed an ideal area to explore. I realized that while Intel had many of the technical skills to develop networking products, it lacked the marketing and business expertise required. It was doubtful that Intel had the time to develop those skills internally before other companies achieved strong positions. I wanted to acquire a company that could jump-start the process, that could form the nucleus of a networking business.

Les was very supportive of this idea. He even arranged for me to make a presentation to the Intel board of directors on networking. I used the automotive industry as an analogy. I explained how the oil companies, like Standard Oil, pushed the development of highway systems by building small demonstration "highways" and providing free maps. In my description, the highway system represented computer networks, while the cars were the computers. I explained that once the highways were built and legislation limited driving speed, the automotive industry found itself no longer competing on speed but by using racing stripes, car seat fabric, and bumpers. To illustrate this point to the board, I even brought a chip upon which I had painted a racing strip.

If we did not want the same thing to happen to us, I told the board, we had to make sure that computer networks were built and that they were fast. Several members of the board at that time understood more about the computer business than did the management of Intel. This included Max Palevsky, who had founded Scientific Data Systems. While none of the board had experience with networking, which was still in an early phase of development, they were supportive of my exploring this area.

The networking industry was very fragmented then, with some companies offering network adapters, others offering hubs and routers, and still others offering software and servers. I believed that Intel could become the leader in networking by rolling up a number of these smaller companies. To accomplish this, though, we would need talent to lead this effort. I concluded that the only way to do so would be through acquisition.

As a novice with respect to mergers and acquisitions, I sought the help of investment banks in New York City, like Goldman Sachs and Morgan Stanley. When visiting their offices, I was treated very well as a representative of Intel, an important public company. I discovered that the bankers employed some of the best chefs in New York to work in their executive dining rooms. This seemed to be a big part of their sales approach. I had almost no prior experience working with investment bankers, besides the ill-fated Franklin IPO. The bankers were very enthusiastic about helping me, as they had a number of interests with respect to Intel. They traded Intel stock for their customers, and they would also earn a great deal in commissions on any acquisition Intel might make. They also wanted to increase their standing with Intel, whose treasury now contained significant assets.

The bankers had analysts that followed various industries and made recommendations to the firm's clients on which stocks to buy or sell. I met with many of them. They helped educate me, not only with respect to potential acquisition candidates in the networking business but also about the technical aspects of how acquisitions worked. There was so much for me to learn. I felt like I was back at school (then again, I never really went to school). In addition to being coached, I read university textbooks on the topic of mergers and acquisitions (M&A). I had to become an expert in high finance in a very short period of time.

I arranged for several of the investment bankers to visit Intel. In presentations to Andy, Les, and Arvind Sodhani, Intel's treasurer, they outlined strategies for Intel to enter the computer-networking business. These presentations were helpful. There was definitely high-level support for acquiring a networking business, in theory.

We could not invite the bankers to Intel's executive dining room, because we didn't have one. In Intel's cafeteria, everyone from Andy Grove on down stood in line and waited. And everyone had to bus their own dishes. We also had no assigned parking spots or, really, any other perks. Sometimes Intel could be egalitarian to the extreme, probably a result of the founders' humble roots. I will never forget Andy telling me that he had a vacation home in the ski area of Lake Tahoe, which he owned together with another Intel executive to save costs. He said he did not really like going up there because he had to do so much maintenance work himself. Andy must have been worth more than $100 million at the time, but it did not occur to him that he could hire people to do that work.

A BIT OF BACKGROUND ON NETWORKING

Before personal computers and local area networks, there was time-sharing. Individuals might use a terminal connected to a mainframe or minicomputer. Terminals normally only had an alphanumeric, monochromatic display. They connected to computers by serial port. The display speed was very slow by today's standards; you could see each character written one at a time on the screen. When personal computers were first introduced into businesses, they used a serial port to connect a PC to the time-sharing computers and ran terminal-emulation software. As more and more PCs appeared within the organization, people sought to find a faster way to connect them to what we would now call servers, developing the concept of a client/server network. Email was a very important factor in driving this.

Initially, computer networking was developed to allow mainframe and minicomputers to communicate with each other. In 1975, for example, to connect PDP-11s together, Digital Equipment developed DECnet, a very powerful set of software protocols that could have become the standard for computer networking had Digital not kept it proprietary. Networked systems included computers that were located near each other, computers that were located within a campus, and, eventually, computers that were located remotely from one another. The Internet evolved from a way to connect computers that were both local and remote. The key to the Internet became the router, a device that can pass information from one computer to another via many paths. Cisco would come to dominate this aspect of networking. Each computer then had to have a way to attach itself to a router.

A number of competing technologies developed and deployed in the early 1980s, called local area networks (LAN), could connect computers and devices located within a relatively small geographical area. One of these technologies, Ethernet, was first developed by Bob Metcalfe at the Xerox Palo Alto Research Center. Later supported by Digital Equipment and Intel, Ethernet came to dominate LAN technology and became the way computers would attach themselves to routers. Servers were developed around the same time to support email and shared printers.

Connectivity across large geographical areas, called wide area networks (WAN), developed much earlier in response to the need to connect mainframe and minicomputers to each other over large distances. Much of the technology now used for Internet communication came out of this. Using this same technology, clusters of local area networks could be connected together. If you have high-speed Internet in your home, for instance, you'll notice that your router has one port labeled "WAN" and a bunch of ports labeled "LAN." This is the same general architecture as was first developed for enterprises. In 1999, Wi-Fi was introduced, with Apple offering the first commercial product. This would prove consequential in the proliferation of computing but did not play a role in my thinking at the time.

— can't afford anything we want and
don't want anything we can afford

Networks consist of computers that utilize networking technologies like Ethernet to connect to other computers both locally and remotely. While Ethernet provided a very effective way for computers located close to each other to communicate, other technologies would be needed to let computers connect both to other local computers but also computers that were located remotely. The router is what makes all this possible. While there were a number of router companies, I was convinced that Cisco, which had been formed in 1984, would be the winner in this new market.

So in 1988, I wanted Intel to buy Cisco, which was still small then. We might have bought them for less than $100 million, but Andy thought that was too expensive. Cisco went public in 1990 at a valuation of $224 million. Sometime in 1994 or 1995, Intel took a look once more at Cisco. Once again, Andy passed at the opportunity, saying that Cisco at $4 billion was too expensive. I then bet Andy that one day Cisco would be worth more than Intel. He laughed at the thought. On May 2, 1998, Cisco passed Intel in market capitalization.

Looking back, I may not have done a good job of explaining why Cisco would be so successful. Most likely, I did not appreciate how little understanding Andy had about networking. That's on me. But it was very hard to "teach" Andy anything, at least in my experience. Les was the opposite. He really wanted to understand, and he had a big appetite for knowledge.

Having been turned down by Andy with respect to Cisco, I then considered 3Com as a possibility. Ethernet provided a way for multiple computers to all connect to each other over the same network and a small geographical area, and a device called the router allowed both local and remote connections. The founder of 3Com was Bob Metcalfe, who had developed Ethernet while at Xerox PARC. I knew Bob from my time at Digital when I gave his new company, 3Com, a very large order for Ethernet cards. Though the leader in Ethernet, they were having issues in their market, which was getting to be very competitive.

Andy thought they were a "piece of crap." When he told me that, I angrily responded, "I got it; you don't want anything we can afford, and we can't afford anything you want." That was pretty much true. Looking back, I don't think Cisco, 3Com, or any other potential acquisition would have entirely worked out. At that time, though, I was not yet aware of the problems of integrating such acquisitions into a company like Intel. I believe Cisco would never have become the company they became under Intel ownership. While Cisco had a strong technical capability, they were really customer focused. 3Com might have been a more successful acquisition, but as part of Intel, it would have had difficulties competing with network leaders like Cisco.

The next company I considered was Sytek, the original broadband company formed in 1980. In 1982, General Instruments (GI), the leading supplier of products to the cable television industry, began a series of key investments in Sytek that led them to own 57 percent of Sytek. Though Sytek had ample opportunities to raise venture capital, they believed a strategic investment from GI made more sense. GI was interested in Sytek's development of MetroNet, one of the company's main efforts and one with which GI could help directly. Sytek and GI's collaboration on MetroNet could have been the foundational technology for interactive television and became one of the forerunners of broadband Internet.

Sytek combined off-the-shelf components that were readily available in the television industry, coupled to a key Ethernet component that Intel had developed. There was a demonstration of this technology in Sacramento, California, in 1982. This was the first metropolitan area network, I believe. Later, in 1990, a Boston-based company, LANcity, would do something similar, which they actually brought to market. Though General Instruments was interested in the residential market, the cost of TV set tops capable of connecting to this network was estimated to exceed $1,500, which made it unviable as a consumer product. Consequently, GI began to lose interest in Sytek. In 1992, Intel and GI would partner, utilizing many of the same concepts to develop the cable modem technology that evolved into what is now utilized by the cable industry for residential broadband.

Fortunately for Sytek, its executives ran into twelve members of the IBM Personal Computer Group at a networking seminar in 1983. By that time, the IBM PC was doing extremely well and was becoming a significant contributor to IBM's profitability. The PC group needed LAN technology to serve their market. IBM was in the process of developing its own proprietary network technology called Token Ring, which was meant to compete with Ethernet. The PC group got tired of waiting for Token to be developed. They decided to use Sytek's technology because it could be used both to connect local networks but also to connect enterprises including building miles apart. IBM made a small investment in Sytek. The PC group began marking Sytek's product, which was called LocalNet 2000. IBM and Sytek jointly developed a critical piece of networking software, NetBIOS. Later, Microsoft would incorporate this into their operating system.

Eventually, internal IBM politics killed the Sytek approach and required their PC group to wait for Token Ring. At the same time, the rest of the PC industry moved to Ethernet, which eventually won out and still is the standard for local area networking.

Sytek was in financial trouble once IBM had terminated their purchase agreement and moved to Token Ring. General Instruments, which no longer had any strategic interest, wanted to sell its 57 percent stake. As a result, the Sytek board decided to sell the whole company and engaged an investment banker to conduct an auction. Intel was contacted by their investment banker in 1989, and we began to evaluate making an offer. Les and I liked Sytek's CEO, George Klaus, and thought he had the potential to lead an Intel networking business. The company had a number of strong technologists who could have helped Intel a great deal.

After many internal meetings, we finally decided to make an offer. I then learned an important lesson about managing the acquisition process. Companies up for sale often hire an investment banker to represent them, and the banker may recommend an auction. The banker prepares an initial presentation on the company and reaches out to a number of potential buyers. Next, they narrow it down to a few possible buyers that appear to be very serious and a good fit. Those companies get an opportunity to review a lot more material about the company for sale and meet with all the key employees. The potential acquirers then place bids. The company to be acquired chooses among those bids based on a number of factors, with financial concerns typically prioritized the highest. This process tends to work well for the seller, as buyers who know they have competition will tend to offer more in order to be successful in their bid.

If the potential buyer has a lot of people involved in the deliberations around an acquisition, there is a good chance that someone will say that the price being considered is too high. Usually, then, others will jump on board to say that the price was too high. Later, if you can get the business for the lower price, they will claim they were right. If you lose the bid, they will say the acquirer overpaid. This happened exactly in the case of Sytek. We lost the bid to Hughes Networking, a division of the Hughes Aircraft Company, which later created DirecTV. Sytek became Hughes LAN Systems. Our bid was $20 million less than the Hughes bid of $87 million. After the acquisition, their focus shifted to using satellites for data communications. Eventually, this part of Hughes, Hughes Communications, was sold to EchoStar.

I found this loss very upsetting. It was my real first attempt at an acquisition. I realized that Intel would probably never successfully bid in an auction because we would always underbid. The process works differently when you reach out to a company that has not been thinking of selling and enter into negotiations.

With the advantage of hindsight, I now think that had we successfully acquired Sytek, the evolution of both the networking business in general and Intel could have proceeded very differently. With Sytek's talent, we would have developed the semiconductors needed for broadband. Coupled with Intel's existing Ethernet capabilities, Intel could have become the leading semiconductor supplier in the broadband market, the position later occupied by Broadcom. Perhaps we would have used the knowledge we gained in the acquisition to develop network server technology, as well. Sytek would have been a better cultural fit with Intel than the later acquisitions we made.

THE ANTIBODIES ATTACK

Having failed to acquire Sytek, we ended up buying a small networking company, Jupiter Technologies in July 1989. Jupiter, a software company, dealt with protocol translation. Before TCP/IP became dominant, different computer systems had varied and incompatible methods of communicating. Jupiter wrote software that translated among these different systems.

What I wanted in Jupiter was the talent to build Intel's networking business. Jupiter's CEO, Jim Flach, moved out from the Boston area to work at Intel in Santa Clara. He was made vice president of the Systems Group, reporting to Les. Then the Intel antibodies came out and attacked the "foreign body." Finance, legal, and human resources all would swarm over the "body" of any acquisition, forcing them to adopt Intel's way of doing things. Intel people would treat individuals within the company being acquired as second-class citizens. Though I could provide some cover to senior executives from the acquired company, I could not alone manage the magnitude of the reaction from the Intel "immune system." I was so naïve. Integrating a new company is a very difficult task that requires dedicated, experienced management that we did not have in place. Jim Flach stayed at Intel for three years before leaving to become a partner at Accel, one of the leading venture capital firms.

In May 1991, we also acquired a division of LAN Systems, a company founded and led by Tyrone Pike, a serial entrepreneur. Tyrone joined Intel but did not stay long. However, Ed Ekstrom, the person who led the division we acquired, stayed for many years and made many contributions. PCEO was already reselling LAN Systems' network diagnostic products.

— the challenge of an acquisition is keeping the talent

Though the LAN Systems acquisition was successful, I was learning how difficult it was to acquire companies within Intel. I had thought that by buying some smaller companies, we could get both critical technical skills and entrepreneurial management, which together would help Intel to create its own networking business. This did not work out as planned. The management of acquired companies generally did not want to be part of Intel. Entrepreneurs by nature greatly preferred to run their own businesses. Ultimately, we did not successfully acquire the senior talent we needed to develop our own networking business.

I still believe that networking was a great opportunity for Intel but did not know how to get it to happen.

Looking back, I have to conclude that to build a new business within Intel, there had to be a champion—a person willing to put their career on the line to lead the effort. Perhaps, if I had told Les and Andy I wanted to start a network business for Intel, they would have agreed. But my experience at Digital and then later at Franklin had convinced me that an operating role was not for me.

ZERO-BILLION-DOLLAR BUSINESSES

When Frank Gill took over the System Group from Vadasz, it already had a small networking business providing Ethernet cards and print servers. Slowly, Frank came to the same conclusion as I had, along with many others: Intel had an opportunity to create a large networking business. Frank wanted to grow the business internally; he was not interested in acquisitions or strategic investment. The Systems Group was renamed the Small Business and Network Group. Eventually, this business did more than a billion dollars in annual sales, but Grove would still not invest in it. He required that the business be self-funding. Andy even taunted Gill, comparing the microprocessor business with Gill's network business: "Frank, you sell a billion dollars in one year. I earn a billion dollars in profit every quarter. Why should we invest in your business?" By the time Andy asked this question, he might have been right. The networking business was maturing, and companies like Cisco were very entrenched.

Intel's revenue and especially its profit were so high that Andy saw a billion dollars as a rounding error. I would sometimes refer to the new business efforts as the zero-billion-dollar businesses. On the other hand, Intel Capital would invest in start-ups that barely had revenue and were sustaining substantial losses and watch them grow into companies with billion-dollar revenues, but none of them did this in their first year.

Early in 1998, Frank Gill had proposed acquiring FORE Systems, which was a communications switch company. Craig Barrett, who had just become CEO, very much supported this acquisition and brought it for approval to his first Intel board meeting as CEO. Andy Grove was now chairman of the board. The cost of the acquisition exceeded $2 billion. In a very unusual move, Gordon Moore, who was serving as vice chairman, spoke against the acquisition. Then the other directors agreed. Frankly, I doubt that many of them even understood the issues. Andy, who had favored the acquisition, just sat the discussion out. Soon after this rejection, Gill was gone. He left Intel soon after Barrett was made CEO, a job that Gill most likely wanted for himself. Mark Christensen, who had reported to Frank, took over, but the networking business slowly died out. While Frank never thought I was a "real man" because I did not run a business, several of my networking investments earned more profit for Intel than Frank made the whole time he ran the Systems Group.

By the end of the '90s, Intel came to recognize that there was a major opportunity to provide the critical silicon or, as Intel could call it, "the building blocks" of the Internet. The company started an acquisition spree starting with Level One Communications, which was purchased for $2.2 billion dollars, followed by the purchase of a Danish company, Giga, for 1.25 billion, which specialized in the chips needed for high-speed optical communications. Intel started building a portfolio of specialized semiconductor companies. It acquired tens of companies. From discussions I had with a number of individuals that were involved, it appears that they overpaid, perhaps did not do an adequate amount of due diligence, and had difficulty with integrating the new companies within Intel. Intel had hoped that its manufacturing strength would allow the company to offer the communications/networking components with a price advantage to other companies. Intel would be selling these components to companies like Cisco, Juniper, and Lucent, who, in turn, sold to network operators like AT&T. Once again, Intel was several steps removed from the end customer. In 2002, Network Operators experienced a severe decline in their sales, and they cut back on their purchase from companies like Cisco, who stopped buying product from Intel. In 2007, Intel sold off its optical communications business to Encore for $85 million dollars.

DOUBLING DOWN

After I left, Intel acquired many companies during this period—more than fifty. It looks to me like they did not really have a strategy and were not very good at due diligence and terrible at integration. Intel probably wasted many billions of dollars. Paul Otellini, who was then CEO, wrote in the 2007 annual report that Intel was divesting itself of this business to focus on markets where Intel CPU architecture gave the company a competitive advantage. Intel had once again failed at establishing a new business. It was a very expensive experiment.

Intel would have quite a successful program to accelerate the adoption of Wi-Fi driving standards, integrating Wi-Fi with processors like Centrino. Intel Capital carved out $150 million for early-stage investments in early-stage companies focusing on Wi-Fi opportunities. Sriram Viswanathan, whom I had trained, led that successful effort. However, while this certainly resulted in an expansion of PC sales and therefore Intel microprocessor sales, Intel did not participate directly in the market with Wi-Fi products.

Intel also started to assemble a number of companies to provide the technologies needed on the client side of residential broadband.

The leader in that market was Broadcom, a company in which Intel had once owned a substantial share of stock. Intel considered buying an Israeli cable modem component company, Libit, which we had invested in during my time at Intel Capital. Texas Instruments acquired that company instead. Then, years later, Intel acquired all of Texas Instruments' cable modem business. It would take another eight years, but Intel would become the leader in broadband components, twenty-five years after we started working on the first cable modem. Then in 2020, Intel would exit the market.

Looking back, I realize that Intel in the early '90s should have acquired Broadcom and Qualcomm when those companies were still small. Those acquisitions, along with the acquisition of a leading graphic chips company like Nvidia, would have created a much stronger platform for Intel's future. In 2021, Qualcomm is worth almost as much as Intel. Nvidia is worth twice as much as Intel. Broadcom was acquired in 2015, and while it would have been a successful acquisition, I think, it would not have been as meaningful as the other two.

MY LAST ACQUISITION ATTEMPT

While I was exploring how to create a significant networking business, I was also trying to help Les grow the System Group's PC business. In 1990, I found a successful UK company, Apricot Computer, in Birmingham, which had built an IBM PS/2-compatible computer. I believe this was the only PS/2-compatible computer ever built. My interest in Apricot was primarily focused on the strength of their technical team. I thought that the acquisition of Apricot would have given us a solid platform for European expansion.

My attempt to acquire Apricot turned into a real disaster. Les had criticized me for not working closely with key internal groups. He was more or less correct, so with Apricot, I worked really hard to get all the key people in the Systems Group to review and approve the deal, particularly the technical people. Then, at a meeting to approve making an offer to buy Apricot, some of the technical people surprised me and started attacking the deal. They were calling in from Oregon while we were in Santa Clara. I couldn't understand why they were doing this; the technical people had all previously told me they supported the deal. Then Andy started in on me, which, believe me, was not fun. To my surprise and great disappointment, Les, rather than defending me, piled it on as well. The deal was rejected. I was very angry with Les for not supporting me. This was one of only two times we had a major conflict. This was a big one; I stopped speaking to him at all for some time.

I did not give up on the deal. Those who know me well can testify to my tenacity. I began discussions with Giorgio Ronchi, the CEO of Memorex Telex, a company selling IBM-compatible disks and terminals. We discussed setting up a joint venture to sell IBM-compatible PCs from Apricot through the Memorex Telex sales force. With Memorex Telex interested, I brought the potential acquisition of Apricot back to Intel. As I expected, having a serious customer for Apricot's products changed Intel's thinking about the acquisition. Still, it took me many months to get the deal approved. Then we had to complete due diligence.

Dick Boucher, to whom I reported while Les was on sabbatical; Arvind Sodhani, Intel's treasurer; and I were the senior team responsible for the due diligence. This was to be my first large-scale acquisition. Every department in Intel—finance, legal, human resources, and security, among others—had to get involved to verify everything Apricot had sent over and look to bring any problems to the surface. Some twenty people came with me to the UK to conduct the due diligence on Apricot. I have no idea how it came about that so many Intel people were involved. In retrospect, I should have made sure I understood in advance the totality of the effort. For instance, since we would be buying assets like buildings, the people that managed Intel's facilities wanted to be involved.

Soon after we started our work at Apricot's headquarters, I realized that I had lost complete control over the process. After three days, the chairman of the board of Apricot Computer, Roger Foster, asked me to come to his office. He was calling off the deal, he said; he would never subject his people to working for a company like Intel. Pointing to his window, Roger showed me what was, for him, the final straw. There we saw an Intel crew digging a big hole in his parking lot, looking for any hidden industrial pollutants: just standard Intel procedure, a step that would have made sense if we were buying a factory that manufactured semiconductors.

Boucher and I collected all the Intel people and returned to the USA. Apricot was the last time I tried to complete this kind of acquisition at Intel. Larger acquisitions of operating companies are very difficult in high tech, although some companies, such as Cisco, have developed skill at doing this. John Chambers, Cisco's CEO, discusses this in depth in his book, *Connecting the Dots*. Acquisitions of smaller companies to get talent and technology can make sense, as Apple, Google, and Microsoft have demonstrated. While I still believed that Intel needed to create new businesses, I no longer saw how I could contribute to making this happen.

10. ADVENTURE CAPITALIST

I was learning painfully that acquisitions would not work at Intel, and it had also become clear to me that joint ventures, such as BiiN, are like running in a three-legged race, requiring lots of coordination and little speed. While strategic partnerships could be useful, they were typically short-lived, other than the symbiotic relationship between Intel and Microsoft. That left me with exploring minority investments as a way to expand Intel's business.

I had a small and inexperienced staff of just two people. At the same time, Harold Hughes, Intel's former treasurer, had moved to the components group, where he was also making a few minority investments. I started focusing on some early-stage investments, including one in GO Corporation at the request of Andy Grove. GO was a company building one of the first personal data assistants. I am not sure why Grove wanted to make that investment. We required that GO use the 386 and that they would never release their software on competitive microprocessors. It was probably an insurance policy in case GO was successful. Andy was not a big believer, correctly, in pen-based computers. The legendary Bill Campbell was the CEO. He and I became friends and that lasted until his death in 2016. The founder, Jerry Kaplan, wrote a very detailed book about GO Computer, called *Startup*, which is worth reading.

I believed that minority-investment opportunities could provide Intel with strategic insight and market impact without raising issues of integration. Les got Andy to agree to allocate an initial $50 million for this purpose. At first, Andy attended the meetings where we discussed potential investments. Frankly, his presence was not constructive. Les was able, after a while, to convince Andy that he probably had better things to do with his time.

In August of 1988, I made my first early-stage investment in Digital FX, led by my friend and co-founder of Atari, Steve Mayer. The hugely successful venture firm Kleiner Perkins also invested, along with Apple, Sevin Rosen, and Mitch Kapor. Before joining Intel, I had helped Mayer spin out Digital FX from Warner Communications, the company that later merged with Time Inc. to form Time Warner. As part of a settlement of Steve's contract, Digital FX got rights to some technology and cash, which was followed by a loan from Apple. Though I had no ownership in Digital FX, I strongly believed in its potential and thought that an investment in that company would help boost Intel's understanding of the importance of video technology to the development of the PC market.

I remember chasing Andy down in the Intel parking lot to discuss the Digital FX investment. He planned to attend the meeting where the investment would be approved. Certain he did not really understand the company's business or its strategic possibilities, I thought it would be best to meet with him to explain before that meeting. Yet, I could not get on his calendar. When I saw him, through a window, walking out to his car, I ran out of the building and grabbed him. I told him that I had been trying to get some time with him to discuss the Digital FX investment. He acknowledged that he had seen the request, but he said that he did not have relevant knowledge to help me better understand the opportunity. In other words, he thought that I wanted him to help me. By contrast, I wanted to make sure he understood the subject matter and would not be embarrassed by his lack of knowledge. Andy was comfortable in the role of teacher but not that of student, at least with me. Somehow, I got through the meeting and the investment was approved.

The initial business of Digital FX was to replace very expensive equipment used in television production to create special effects, titling, and so on. The idea was to use standard PCs with an add-on board to compete with systems that were four or five times more costly. The use of PCs with add-on hardware and specialized software fit Intel's strategy, but my interest went further. I thought we could eventually significantly reduce the cost of the hardware by using specialized silicon, making a consumer product possible. I envisioned a layer of software above MS-DOS using Digital FX's technology to provide a version of Windows with much better performance than PCs at that time could offer. We could start with an add-in board before eventually integrating these capabilities into the main microprocessor.

For a while, Digital FX did very well, even winning an Emmy, but it eventually ran into financial difficulties and closed. The consumer product I had imagined was never developed. Intel would later add critical multimedia capabilities to its microprocessors, but Digital FX technology did not play a role.

LES GOES ON SABBATICAL, AND I ALMOST GET FIRED

Les was having difficulties as head of the Systems Group. One year, Andy wanted the group to be more aggressive; the next year, he thought they should back down because they were competing too aggressively with Intel's main customers (PC manufacturers). The management team of the Systems Group was in disarray. Andy decided to change its leadership.

Les took an extended sabbatical to attend the Harvard Advanced Management Program and then briefly taught at Harvard. With Les gone temporarily, Frank Gill, formerly the head of the Intel Sales Organization, took over the Systems Group. Frank and I had a good personal relationship, but now that he was my boss, he had no use for my investment activities. It seemed that he was highly suspicious of my plan to make money through minority investments and also not interested in acquisitions. In Frank's mind, I suspect, "real men run real businesses." He did not want to supervise my work and made it clear that I needed to find a new boss if I wanted to stay at Intel. Luckily, another senior member of Andy Grove's staff, Dick Boucher, bailed me out. Dick and I had cubicles next to each other and were good friends.

While he did not want any direct reports, he had a lot of respect and affection for me and understood that he had to save me from Frank until Les returned. Dick was one of the nicest and kindest people I knew at Intel. We are still in touch.

INTEL CAPITAL (CBD) IS BORN

Les returned to Intel in 1991 after being at Harvard for about six months. Andy asked him to focus primarily on business-development activities, including venture investments. While Andy thought our merger-and-acquisition and minority-investment activity made a lot of sense, he told Les he felt this activity needed senior management oversight if it were to expand. Les was to provide that oversight. Andy putting a very senior executive like Les in charge was critical to the future success of Intel Capital. Les named our new group Corporate Business Development, or CBD. It would be renamed Intel Capital in 1999, the name I will use from here on to avoid confusion. Without Les, there would have been no Intel Capital.

The work I had done in business development from 1988 until Les returned from his sabbatical in 1991 formed the basis for the creation of Intel Capital. For that reason, and with his blessing, I call myself a co-founder of Intel Capital.

My title was now vice president of corporate business development. For almost six months, I was Les's only direct report. We had a great relationship and felt like partners, with Les the senior partner. Indeed, a February 2000 *Wall Street Journal* profile of Les gave me the honorable epithet, "Vadasz's Lieutenant."

Intel Capital, I believe, represents one of Intel's most successful undertakings, after the development of the microprocessor itself. By the time I left Intel in 1999, we had made more than five hundred investments, and the portfolio was worth an excess of $8 billion. At its peak in 2000, its value reached $15 billion, not counting $5 billion in gains already realized. Most importantly, Intel Capital had a major impact in terms of growing the market for Intel's products, itself worth tens of billions in additional Intel profit. In 1999, more than a hundred professionals worked at Intel Capital, and their number may have doubled in the early 2000s. Unfortunately, many of the unrealized gains were wiped out when the Internet bubble burst. Still, Intel Capital was one of the largest venture capital organizations in the world, if not the very largest. Venture industry metrics still, thirty years after we formed it, place Intel Capital as one of the most successful, most active corporate venture groups in high tech.

STRATEGIC OBJECTIVES FOR INTEL CAPITAL

Les and I established three major objectives for Intel Capital:

- Help shape Intel's long-term strategy
- Grow Intel's current business
- Get a reasonable financial return

We thought of ourselves not as venture capitalists (VCs) but as an organization tasked with expanding Intel's business, which is why we originally chose the name Corporate Business Development. Neither Les nor I ever identified with the label "VC." For Intel Capital, venture investing was a tactic to implement our larger strategy. Indeed, my role throughout my later career at Intel was broader than venture investing, including participating in the strategic long-range planning process; reaching out to the entertainment, media, and telecommunications industries; acting as spokesperson; and even testifying at congressional hearings.

Avram testifies at a congressional subcommittee on communication in 1998.

I take more pride in my efforts to develop the consumer Internet, particularly to develop residential broadband, than I do in the billions of dollars we earned on our investments in that specific area. Although I am indeed proud of the financial returns from our venture investments, even taken specifically in those terms, Intel Capital's efforts to grow the market for Intel's products and to make the PC the dominant interactive device in the home had a much greater impact on Intel's financial performance than our considerable returns.

While Intel Capital did have some influence on Intel's long-term strategy, primarily by giving the company insight into future market and technological developments, both Les and I are disappointed that the impact was not greater. While an internal venture group may provide important strategic insight, it cannot cause the actions needed to take advantage of this insight. Perhaps we could have done better at communicating what we were learning. I know we tried.

At first, as strategic investors, we planned to earn reasonable but not outsized investment returns. For instance, we might be willing to invest in a company that would not normally receive venture capital because they could be helpful in developing markets for Intel. We soon began to realize that the more successful portfolio companies were, the more likely they would have the impact we sought on the broader industry. We also required that we be joined by lead investors who had only financial interests, and they would set the financial terms.

Because of Intel's central role in the computer industry, we had a great deal of insight into what the various leading companies were doing and planning. While we couldn't share proprietary information with our portfolio companies or partnered venture capital firms, such information did greatly shape our own investment decisions. And just an announcement alone of any investment by Intel could prove hugely valuable to a portfolio company. Achieving great returns right from the beginning gave us a lot of independence and freedom within the corporation, since we generated the cash we needed for further investments many times over.

When we started, neither Les nor I had much experience in venture investing, but we fortunately did have very complementary skills. He understood the benefit of clear, effective processes and knew how to get things done within Intel. I, on the other hand, could see the market and technology trends that might be relevant to Intel and felt comfortable building relationships outside of the company.

Les knew that all aspects of Intel's business could benefit from strategic investments. For instance, he wanted to invest in semiconductor process technology and hired someone to manage this activity, reporting directly to him. Another area where Les was aggressive was enterprise computing. As Intel got more comfortable with Intel Capital, more business units came looking for help making strategic investments. Les hired someone to handle enterprise computing. We expanded internationally, with people in Europe, Israel, and Asia.

I began to grow my own group as part of Intel Capital and was able to hire some talented people. I am still in touch with most of them, many of whom went on to have very successful careers in venture capital.

THE METHODOLOGY

We wanted to make sure that nothing we did would reflect poorly on Intel, an issue that became more significant as Intel Capital grew and involved increasing numbers of people. We wanted to enable our "deal-makers" to negotiate with prospective companies while also ensuring that the deals they ultimately proposed would be acceptable.

We devised a two-step process. In the first step, called the deal concept meeting (DCM), the Intel Capital dealmaker presented a potential company for investment and discussed the general terms of the deal they wanted to explore, including the amount of potential investment and expected pre-money valuation. We were rarely the lead investor, preferring to have professional VCs set the valuation. The VCs sometimes brought us deals, and sometimes we brought them deals. At first, all investments had to have business unit sponsors. In the beginning, we viewed our role as growing Intel's business by helping the business units expand. I felt this was too limiting as the time frames of the business units were much shorter than the time frames for early-stage investments. This was especially true for the home-computer market where there was not even a relevant business group. Les agreed with my request to no longer have to have business units support my investments given the time horizon.

The next stage was the investment project approval (IPA). In this meeting, the terms of the deal were discussed, along with clearly stating Intel's strategic objectives. Typically, deals that got to the IPA were approved. The time between DCM and IPA would usually be thirty to sixty days. There were, of course, situations where things moved much faster. For example, we invested in CNET just before its IPO. We only had a few days to make it happen. For many years, Les was the only person authorized to approve an investment. Eventually, I was also given that authority as well.

Les wanted active involvement from both the legal and finance organizations. He was able to get these organizations to provide dedicated resources in support of our effort. The legal and finance representatives practically had veto rights on a deal. The dealmaker needed to make sure they were on board and resolve any of their issues. The legal department would make sure that the deal was structured correctly to protect Intel's interests. The finance department would take a position as to the valuation and the terms. This resulted in a three-person team for every deal. It worked extremely well. I can only remember a few times out of hundreds of deals when there were disputes. Suzan Miller was responsible for the legal group supporting Intel Capital, and Randy Tinsley was responsible for the financial group supporting Intel Capital for most of the time I was part of the group.

Intel Capital also started having yearly meetings where all the portfolio companies were invited. We promoted interaction between them and also invited key Intel executives to attend in order to give them exposure to the companies.

We worked closely with several venture firms. We would provide each other with deal flow and support the due-diligence work needed to make an investment determination. We preferred that the VCs would take the lead investor role.

While we worked with many firms, I was particularly close to Accel Partners (Jim Breyer), Sequoia (Doug Leone), and Redpoint (Geoff Yang). We also had relationships with other corporate venture groups such as General Electric, Lucent, and General Instruments. While Microsoft was our closest business partner, we did not typically coinvest with them. We also occasionally coinvested with Kleiner Perkins, but we did not have a close relationship, and frankly, we were not that comfortable with them. Generally, we would find the deals and introduce the companies to the VCs. We did not take board seats because we were concerned with creating liabilities for Intel. However, we would often get the right to send observers to the board. I would go to the board meetings of the larger investment we made, like CMGI, while members of my staff would attend the board meetings of some of the smaller companies. This gave important training to all of us.

THE ROLE OF CORPORATE VENTURE

Traditional venture capital firms have one primary mission, which is to generate financial returns for their limited partners. They usually do this by investing small amounts of capital in many companies. They invest additional amounts, usually at higher valuations, in companies that look like they have improved their chances of success. VCs may specialize in a certain sector, like health care, consumer products, or telecommunications. They may also focus on various stages, like seed round, Series A, or Series B, for instance.

VCs raise "funds." Most investments are institutional, such as pension funds. VCs generally get a 2 percent management fee, and 20 percent carry interest (20 percent of the profits). They typically get liquidity when the companies they invest in are acquired or go public.

Most early-stage businesses do not succeed. But maybe 10 or 20 percent have major successes and return more than ten times the investment.

Some corporations have captive venture capital "firms." These groups are usefully part of a company's finance organization. Generally, I think it does not make sense for corporations to engage in this kind of venture investing.

CORPORATE
Investment Objective

		Strategic	Financial
Link to Operational Capability	Tight	**Driving** advances strategy of current business	**Emergent** allows exploration of potential new business
	Loose	**Enabling** complements strategy of current business	**Passive** provides financial returns only

(From original chart by Henry Chesbrough)

Henry Chesbrough wrote an excellent piece in 2002 in the *Harvard Business Review* on corporate venture. He discusses Intel Capital a bit. He created a four-quadrant model (after all, he is from Harvard). It explores corporate investment's relationship to financial and strategic returns as well as the link to operational capabilities.

Intel Capital stayed pretty much on the strategic side. Most of the work of the group was enabling current businesses. Most of my work was advancing the strategies of the current business. We did some investing in what is called emergent, but not enough. During my time, we stayed away from passive investments. Later, Intel Capital moved more and more to the financial return side, which I consider a mistake.

FINALLY GETTING TO WORK ON
THE CONSUMER SPACE

Intel provided a sabbatical to all employees at the end of every seven years. I became eligible for my first one in August 1991. The sabbatical was two months long. I combined mine with the end-of-the-year holidays, so I actually had two and half months. It was a very interesting time for me personally. The only time when I had such free time was the four months between leaving Franklin Computer and joining Intel, but I had spent that time looking for a new position and was very stressed. My wife, Arianne, used this time to go back to Holland to do research for her PhD thesis. I decided not to travel, and during these months, I exercised two hours a day, six days a week, with a trainer. It was pretty wonderful. Finally, I was strong. Prior to this, I thought of my body as just a socket for my brain. Now one sickly kid was a bodybuilder. I have stayed fit since that time. I also used this time to practice piano about five hours a day, including having three lessons a week.

I was always interested in consumer products. My first experience with electronics was repairing our family's television when I was about ten years old. Later, when I lived in Paris in 1965, my friend and I developed a few devices that were consumer oriented. When I was at the Thoraxcenter from 1969 to 1975, I developed a device to light a light when my deaf sister-in-law and her husband were asleep and their baby was crying. During my first Intel sabbatical, I got a Mac and started playing around with various creative applications. So maybe all of this had an effect on me, or maybe I just no longer liked the business market.

When I returned, I made a pitch to Les and to Andy about wanting to work on the consumer market for PCs. Andy took the position that there was no home market for computers. We argued about that for a while. Then he said, "If you want to waste your time on this, you can, but don't involve anybody else." Later, at a meeting for about two hundred senior managers, Andy mentioned that I was going to be looking into the home market for computers, which he thought was a waste of time, and that no one else should get involved. I then started to spread the word that every Tuesday I was going to be in a certain conference room having my lunch and thinking about the home computer market. People started dropping in. Soon, I was booking a larger conference room. I think all the discussions we had played a role in determining how to position the Pentium when it was launched in March 1993, with the emphasis on the consumer market. Even Les did not think the home market for computers would be that big. I remember him calling me out for saying during a speech that I had given around 1993 that half the homes in the United States would have computers by the end of the decade. He thought this was a major exaggeration. Well, by 2000, more than 51 percent of US homes had a computer.

By 1992, I had realized that the combination of home computers and high-speed residential networks would become a new medium for communication, commerce, education, and entertainment. This realization was one of the most powerful experiences I have ever had. I had observed the development of networking within the office environment, witnessing the transformation of the PC from a computational device to a communications device with the advent of email. Online services like AOL, Prodigy, and CompuServe were becoming popular in homes that had computers. It was clear to me what was missing. That was a reliable, always-on, high-speed capability to connect to a number of services.

I had funded the Intel Architectural Labs to develop cable and DSL modems. The development of residential broadband became one of my primary activities. I led Intel's efforts across various groups. I wanted to invest in every aspect of what it would take to make that happen, and it paid off, generating in billions and billions of dollars in profit for Intel.

As the Internet took off, the value of our investments increased significantly. My group alone was sitting on billions of dollars of unrealized gains. We invested very early in companies like Broadcast.com, Verisign, Broadcom, Launch Media, GeoCities, CNET, eToys, @Home, and Covad. Later, we invested in CMGI, PCCW, and World Online. Of course, we also had our share of failures, such as American Cybercast and 911 Music. The losses did not cost us much, and the winners made up for the losers by more than a thousand times. We also unfortunately passed on several companies that became very successful, the most notable being Netscape. I had become friendly with the founder, Jim Clark. On the day he started up Netscape, he asked me to come to his office where he introduced me to Marc Andreessen, the developer of the first successful consumer Internet browser, Mosaic. They told me about the company they were starting, which was first called Mosaic but later changed to Netscape because of some legal issues. Jim offered Intel the opportunity to invest. The company would have a valuation of $18 million before the investments. I discussed it with Les, and we both thought it was too expensive. We viewed the browser as a tool, and in general, tools do not make a lot of money, so we turned down the offer. A year later, the company went public, and its shares were worth $3 billion dollars on its first trading day. It eventually was worth about $10 billion. Even though I said no to the investment, Jim asked me to provide a quote for the announcement of his company. I am the only non-Netscape person to be quoted in the announcement.

My part of Intel Capital grew substantially, especially after the need to get business unit buy-in was removed from my investments. I evolved a specific approach to investing that was very successful. It started with a concept I called "neighborhoods." I thought if you could identify a neighborhood that would develop, the probability of any property in that neighborhood being successful was greatly enhanced. For instance, we identified e-commerce as a great neighborhood. Another example is the digital distribution of media. These are all obvious now but it wasn't so clear in 1992 when I developed this approach.

Another important concept was what I called "the investment stack." I realized that there would be a number of critical technologies that had to be present for a particular application to be successful. For instance, we first started to explore e-commerce in 1994, the same year Amazon was founded. At that time, consumers were concerned about using their credit cards on the Internet, even though the same people would hand their cards over to waiters at a restaurant. We started to work on making the Internet more secure for consumers. This included an investment in Verisign when it was created to provide for secure transmissions across the Internet. Not only did it make sense to start with the bottom of the stack because that enabling technology would have to be in place for an e-commerce investment to be successful, but those companies were very helpful to us in creating deal flow and validating other investments higher up on the stack. Another example is Broadcom, which developed the semiconductors needed for cable modems. We invested in Broadcom well in advance of investments in cable modem companies and broadband ISPs such as @Home because the components were needed before there could be devices.

— a bad house in a good neighborhood is bet-
ter than a great house in a bad neighborhood

I encouraged my direct reports, or, better said, required them to develop serious expertise in the areas they managed, or their "neighborhoods." One became an expert in e-commerce, while another developed expertise in content distribution, and still another focused on education.

My ability to spot the "neighborhoods" was excellent, it turns out. First, I relied on my intuition to identify them. Then I would develop my own expertise. Sometimes this required hiring an expert who could teach me. I loved learning, so this part of my work was a real joy. I would go very deep in learning about a new market. Then I would go to industry conferences in each area to learn more about what was going on. I would find a way to develop relationships with leaders in various fields. In general, I was working about two years ahead of my people in trying to develop an understanding of future opportunities. This is where my being an extrovert paid off. I socialized with executives and technologists from many different industries.

Launch Media is a good example. In 1996, Matthew Cowan, who was responsible for entertainment investments along with Sriram Viswanathan, introduced me to the late David Goldberg (the former husband of the COO of Facebook, Sheryl Sandberg). Goldberg had started a company to develop what was called enhanced CDs. In this case, it was music CDs that had extra content that you could access from your computer. This was just when residential broadband was being introduced. I told Dave that we would invest if he moved his business to distribute on the Internet. Dave actually did not know anything about the Internet, but he was willing to take the chance, especially since we'd made a serious investment in his company. Yahoo would eventually buy Launch Media, which became Yahoo Music. Another prominent example in this space was our investment in Mark Cuban's Broadcast.com. I had to laugh at the role reversal when the television show *Shark Tank* first aired, because Mark Cuban had pitched his company to me.

While VCs are typically very happy to get ten times return on a few investments, we exceeded this significantly. I was not the only one in Intel Capital getting such amazing returns. Intel Capital was returning billions of dollars of profit to Intel. Several of the top VCs also got fantastic returns during this time, more than fifty times the investment capital. For a while, I think we did even better.

In 1995, I was elected as a corporate vice president. This was a big deal and something I never thought I would achieve. At the time, there were only twenty-seven corporate officers (only two of which were women) even though Intel had more than sixty thousand employees. Les made it happen, but it would never have occurred without Andy's support. I was actually in shock when Les told me. Intel was one of the most successful companies in the world. Intel Capital, an organization with less than a hundred employees, now had two corporate officers, Les and me. Intel was very stingy at the time in electing officers. It still only has thirty-eight officers. Many groups had appointed officers, but they were VPs in name only. The board had to elect corporate officers. Being elected VP increased my status both inside and outside of Intel. It also had a significant impact on my compensation.

— exits are necessary but not easy

DIVESTING

I had a quarterly meeting with my reports to decide on what shares of the companies that had gone public we would sell. This was an essential part of the process, as no one wanted to sell, and I always had to put a bit of pressure on my team to liquidate some of our holdings.

The valuations of the tech sector had driven the value of the portfolio. As a result, the valuations must have been severely impacted by the crash of 2000. In 1999, everyone in tech thought they were geniuses. Then the crash that began at the end of Q1 2000 wiped out most, if not all, of the gains. I was no longer at Intel and do not know how Intel actually handled this period. It was time to increase investment and make significant acquisitions because the valuations of most companies had been severely impacted. It was a significant opportunity for Intel, but not one that was taken. I like to imagine that had I stayed at Intel, I would have done this, but I don't know. When things are going up, it is hard to imagine they will ever go down, and when they are going down, it is hard to believe they will ever go up.

Fortunately, for me, I left Intel in 1999. I thought tech stock values were crazy, but my mantra was "just give me one more year." I wanted to invest for my own account and was able to double my net worth in just twelve months. This was right before the tech bubble popped in at the end of Q1 2000. Just before that, I liquidated all my high-tech public stock holdings, including Intel. Now, twenty years later, Intel's stock price is still significantly below the price it was when I sold my shares in Q1 2000 within a dollar of the all-time high.

It is essential to be disciplined about selling. For me, one of the first rules of investing is, only bet the house's money, meaning, take back your investment as soon as possible. Many of the companies we invested in did not make it ultimately, although they did have a strategic impact like @Home and Covad. Others, PCCW, CMGI, and World

Online, raised significant cash before the crash and survived but had limited long-term impact on the industry.

THE ONE-MAN THINK TANK

Most of the senior management of Intel were introverts. The company's public relations department was desperate for someone with a "user-friendly" personality. I would fill that role. I started speaking at various financial and industry conferences. When a reporter wanted to quote an Intel executive, it was often me. This enabled me to develop strong relationships with many journalists, some of which continue to this day. While my views were often ignored within Intel, I realized that if I could get a journalist at a newspaper like the *Wall Street Journal* or at a magazine like *Fortune* to echo something I thought was important, I could reference that article back at Intel. For instance, I would say to Andy, "Did you see the *New York Times* article about the growth of the home computer market?" For a long time, Andy would argue with me about the home computer market. He did not think it would ever be a big market and did not want Intel to waste time on it. But seeing it mentioned in the *New York Times* got his attention. Of course, he did not realize I was the source.

But all this press attention started to affect my relationship with other Intel executives. It was frustrating to me. It seemed they were jealous of the attention and did not see the benefit to Intel. While I have to admit that I really enjoyed the spotlight and learned to become an effective communicator and loved to speak at conferences, it also allowed me to play an important role in establishing the PC as the interactive device in the home. After a while, there were so many demands on my time that I hired Lakshmi Pratury, who was doing marketing at Intel, to manage my speaking and press activities. Lakshmi went back to her native India later where she was the cohost of TED India and later formed her own conference in India, INK, which is similar to TED.

On March 14, 1996, the newspaper *USA Today* did a one-and-a-half-page profile on me with the unfortunate title: "For Intel, he is a one-man think tank." For those of you who have dealt with journalists, you will know that the headline is written by another person called "the headline writer." I had nothing to do with it. Much of the article was positive about what Intel was doing, but there was too much focus on me personally. A lot of executives were angry about the article, and it fell to Les to deal with me. He asked me to go with him into a conference room where he really chewed me out. I did not know what to say. I certainly had not gone to *USA Today* on my own. It was the Intel PR organization that made a proposal to the newspaper to do an interview with me. The things that were written had already appeared in other articles.

For a while, Les would not allow me to speak with the press without his explicit agreement, but it was not long before he released me from this requirement. But I had learned a lesson and tempered my use of the pronoun "I."

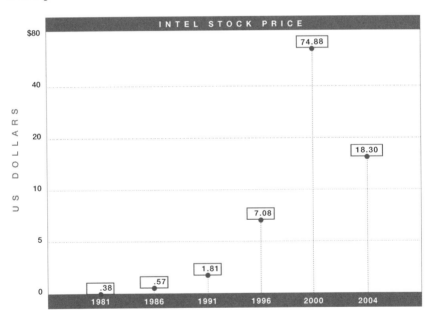

Intel stock price peaks in 2000 and has never recovered.

TECHNICIAN TO THE STARS

The positive media coverage of Intel Capital also brought me to the attention of some Hollywood players. In spring 1995, I got a call from Michael Ovitz. At the time, frankly, I had no idea who he was. As he explained, he was CEO of Creative Artists Agency (CAA), one of the world's leading talent agencies. I had never spoken to a talent agent before!

Ovitz told me that an investment banking firm, Allen & Company, held an event every summer in Sun Valley, Idaho, to which they invited some of the leading CEOs in media, communications, and other consumer companies, as well as political leaders. In essence, it was summer camp for a few of the most powerful people in the world and their families. Ovitz said that Herbert Allen wanted to extend an invitation to Andy Grove and me to attend that year's event, held right after the July Fourth holiday. The first Allen & Company Sun Valley Conference was held in 1983 and still continues thirty-five years later.

I jumped at the opportunity. I was convinced we were creating a new medium for commerce, communications, educations, and entertainment by combining powerful home computers with high-speed connections to the Internet. While my group at Intel Capital was actively investing in early-stage companies for the first three of these, we had found little in the way of new investment opportunities for entertainment besides sports and media distribution. I wrongly thought that the entertainment of the future would come from the same people who had made movies and television shows. Just as so many made the mistake of thinking that the television would be the interactive device in the home, I mistakenly thought that the people who had mastered the creation of passive, linear media would also create interactive media. In retrospect, this was like thinking that the stars of silent movies would also star in talkies.

Andy Grove goes to Hollywood.

The opportunity to go to the Sun Valley Conference with Grove, as I saw it, would let me do two things at once. Not only could I make contact with key entertainment executives and creative people, I might also garner increased support from Andy for my idea, about which he was pretty skeptical, that we would connect millions of computers in the home to online services via the cable network. I told Ovitz that I would discuss it with Andy and get back to him. Then, of course, I did my research on Michael Ovitz, CAA, Allen & Company, and, especially, the conference itself. When I presented the opportunity to Andy, I was surprised and delighted when he agreed to attend.

Herbert Allen's office made the arrangements, which included sending a private airplane to pick up Andy; his wife, Eva; my girlfriend, Harriet Rubin; and me from the San Jose airport. When we landed in Sun Valley, we were confronted with literally tens of private jets, including several of the ultra-desirable Gulfstream IVs.

A few years earlier, Arianne and I had decided to end our marriage but not our friendship or our family. She and I are still very close. Soon afterward, I began a relationship with Harriet Rubin, who lived in New York. She created and ran an imprint for the publisher Doubleday that published books written by some of the most successful businesspeople in the world. It was through our relationship that Harriet got to know Andy well. At the Sun Valley Conference, she was able to persuade Andy to write the very successful book *Only the Paranoid Survive*, published in 1996. I thought it might be awkward for Harriet to publish and edit Andy's book, but he didn't seem to mind. Andy also took on Rob Siegel, who had worked for me as his research assistant. Harriet was impressed by Andy's confidence and clarity, she confided in me. She told me that he was a standout even among the CEOs and entrepreneurs that she worked with. "I had a sense that his mind was open, that he craved the chance to learn things he didn't know and that he had a true North Star, which was that perfection was his life," said Harriet. I don't exactly share Harriet's opinion about Andy.

For the conference, Allen & Company took over the Sun Valley Lodge, as well as the numerous condos that were connected to it. At my first conference, we were given a nice room in the hotel. After that, I was "upgraded" to a condo. They even put a piano in mine.

All the attendees received binders holding the agenda of the various events, as well as a list of the attendees. I so wish I had kept those binders, because now I can only rely on my memory. We also got T-shirts, vests, and other goodies. One of the incredibly amusing things was to see all the billionaires standing in line to receive swag bags from some of the companies.

I remember a few of those in attendance at my first conference. The CEO of Coca-Cola was there, which was not surprising, since Herb Allen was on Coke's board, as was the CEO of McDonald's and Wayne Huizenga, CEO of Blockbuster Video. Wayne had made his money in waste management and later acquired Blockbuster, building Blockbuster into the major corporation it was at that time. Michael Eisner, the CEO of Disney, attended, as did, of course, Michael Ovitz. Nike CEO Phil Knight was there. A number of cable executives were present, including John Malone and Ralph and Brian Roberts. I discuss my interactions with the cable executives in greater detail later, in part III: New Horizons.

For sure, Warren Buffett attended, because I remember going white-water rafting with him. Bill Gates was there, along with his new wife, Melinda, at that time still a product manager at Microsoft. I remember that she gave some kind of demonstration. Bill and Melinda attended the conference every one of the six years I attended. In 1996, they had their newly born daughter, Jennifer, with them. She was about three months old. It took all I could muster to resist saying, "Let me hold the richest baby in the world." Of course, Gates has since decided to leave almost all of his wealth to charity, not to his children, so I'm glad I resisted the temptation.

I remember overhearing an interesting conversation about Paul Allen's new yacht. Someone asked him if he planned to install a heliport. Paul looked at her like she was crazy. "Of course," Paul said, "where else would you put your helicopter?" In fact, his yacht had two heliports.

I had the honor of meeting Katharine Graham, publisher of the *Washington Post*, and her son Donald. I met Mike Bloomberg. I met Steven Spielberg, David Geffen, and Jeffrey Katzenberg, who had recently formed DreamWorks. Andy and Eva seemed to enjoy hanging out with all of these people, which made me happy. Les also attended some of the conferences, as well as Ron Whittier.

In one of the great honors of my life, I introduced Spielberg, the man who established a foundation to remember the Shoah, to Andy Grove, a man who had survived it. Andy often teased me with the nickname "star fucker" because I hung out with people in Hollywood, although he seemed pleased to take a meeting with them. Once I noticed that a photo taken the day Andy and Steven met was taped to the refrigerator in Andy's home.

Andy Grove, me, and Barry Diller

Some of the other notable people I met either at that first meeting or in future meetings included Rupert Murdoch, Bob Iger, Haim Saban, and Barry Diller. Over the years, I met Oprah, Jane Fonda, and songwriter Carole Bayer Sager. The various years kind of run together in my mind. I know I met Carol in 1996, because I have a letter from her soon after suggesting we meet in Los Angeles sometime. I did meet her and her husband Bob Daly (at that time running Warner Brothers) in their Los Angeles home. It was the first time I had visited anyone who had a guardhouse. Many of the media people I met were very interested in the potential impact of computer technology, especially the Internet, on their business. I was "user friendly" and tried to help them understand the technology's possibilities.

CREATIVE ARTISTS AGENCY

After attending the Sun Valley Conference for a few years, I started spending time at CAA in Beverly Hills, hoping to better understand how I could get the entertainment industry to develop content for the PC. Sandy Climan, a very senior member of Ovitz's staff, decided that we should set up a showcase for PC technology at CAA's headquarters, which we called the Media Lab. I hired Sriram Viswanathan, an Intel employee who was already involved with demonstrating technology, to relocate to Los Angles and be responsible for creating and running the Media Lab. Sriram did an amazing job of creating this showcase for the connected PC. At the same time, I sent Matthew Cowan, who worked for me doing early-stage investments, to join Sriram and focus on investments in early-stage media companies. The Media Lab opened on December 12, 1996, in an event that was well attended by many of CAA's clients.

By then, Mike Ovitz had already left CAA, the agency he had created, to become president of Disney. Sandy Climan left as well, going to work at Universal. The team that took over CAA remained committed to our endeavor. Hassan Miah was our main contact then at CAA.

We focused on developing relationships with the "talent" and not so much with the studios, which we felt would have a lot of inertia. We had many discussions with actors, screenwriters, and directors. But not much actually came out of this in the way of new forms of content. The major area where personal computers and the Internet had a significant impact was in the distribution of media. Our investments in Broadcast.com and Launch Media worked out well. We invested in Intertainer, founded by Jonathan Taplin, which was the precursor to Netflix; experimented with interactive entertainment with songwriter Allee Willis; and invested in an interactive soap opera American Cybercast, but they did not work out. I had made the mistake of thinking that masters of linear storytelling could migrate to interactivity. Today, twenty-five years later, with the exception of computer games, there really is not much in the way of interactive entertainment.

While our work with CAA got us lots of publicity, it did not result in new content for the PC. Some of our several investments in early-stage entertainment companies worked out well, while others failed (as is the nature of early-stage investing). For me, personally, working with CAA was a very special experience. I developed relationships not only with CAA but also with the other major talent agencies. I became friendly with Steven Spielberg and Francis Ford Coppola. Robert De Niro visited me at Intel. We were all interested in how the PC and broadband could develop into an entertainment platform. That was not to be. Just as the TV would not be the interactive device in the home, the PC would not replace the television—at least, not for several decades. In 2021, many do use a PC or mobile phone or tablet as their only video device.

My contacts through Sun Valley also almost got me involved in the production of the HBO show *Project Greenlight*. Sometime in late 1998, I think, a friend asked me to meet with actor Ben Affleck and some others to discuss an "opportunity." I was frequently in Los Angeles in those days, so I agreed because I had a great deal of respect for my friend.

We met at a conference room in Beverly Hills. The meeting was scheduled for two p.m. I was there on time, and so were the other people working with Affleck, but he was not there. They started explaining their concept to me. The idea was that first-time filmmakers would be given a chance to direct a feature film. In addition, though, they wanted to do something similar for start-ups, somewhat in the vein of the now successful television show *Shark Tank*.

Around 2:30, Affleck showed up, sweating profusely and with a big bag of food from Taco Bell that stank up the whole room. He began to eat, explaining he had been running behind all day and hadn't had time for lunch. Matt Damon was his partner in this endeavor, along with Sean Bailey and Chris Moore. By this time, I was pretty aggravated and sorry I had ever agreed to the meeting. They started pitching me to have Intel invest. I asked Affleck how much money he and Matt Damon were planning to invest. "None," he said, since they were contributing their name and connections. Incredulous, I asked to confirm, "You are not investing any money?"

"Yes," he said. "Then this meeting is over," I replied. I said my goodbyes and left. *Project Greenlight* did end up on HBO. The start-up part never happened.

MIXING BUSINESS WITH PLEASURE

The Sun Valley Conference was a very relaxed environment. It was pretty unusual for someone in the audience at Sun Valley to bring up controversial issues, but then again, Andy Grove was an unusual guy.

At the 1996 conference, during a talk by the CEO of McDonald's, Andy, who had recently been treated for prostate cancer, went after him in public about the health issues of fast food. At the 1997 conference, Andy attacked Brian Roberts, the Comcast CEO, when Brian was giving a presentation. Andy said that the cable industry would never really effectively deploy broadband. A lot of financial analysts were in attendance, and the financial press picked up the story. His remarks very negatively impacted the cable companies' stocks. Brian was quite upset. We had a good personal and working relationship at the time. He took me aside and asked me to explain what Andy was trying to accomplish. All I could do was apologize; I was just as surprised as Brian. I later had a pretty heated conversation with Andy about it, since I was working so hard at the time to get the cable industry to deploy broadband. I told Andy I worried that his tirade had significantly reduced my effectiveness. In truth, Andy was at best on the fence about cable, thinking the telecom industry would lead the way.

Apart from a few instances of Andy acting out, Sun Valley was great for Intel and great for me personally. Professionally, it gave me the chance to meet some of the most important people in the media industry. I became pretty good friends with conference sponsor Herb Allen and would often visit him whenever I was in New York, at the company's offices. On one occasion, we had dinner together. He is a very special man, and it was a great privilege to know him.

Peter Barton and me

I also developed a few close friendships, in particular with Peter Barton, who ran Liberty Media for John Malone. Peter was a really amazing man. Five years younger than me, he was thin, athletic, and intense. He loved to play jazz piano, as did I, and that formed part of the basis of our friendship. I stayed at his Denver-area home whenever I visited John Malone and others at TCI. Peter became very impressed with my investment strategy. I think he saw in me another version of himself.

He suggested many times that I set up a separate company, owned largely by Intel, that would invest in opportunities based on the PC and Intel's position in the computer industry, a structure similar to what he was doing at Liberty Media for TCI. I told him that it would never happen at Intel. At the 1996 Sun Valley Conference, Peter confronted Andy with the idea in an unusually aggressive manner, rather out of character for Peter. Andy was polite but surprised. I think he thought I put Peter up to it, but I had not. I knew it would never happen.

One evening in 1997, I heard a knock on the door of my Palo Alto home. That was pretty strange. People were not in the habit of just stopping by. Opening the door, I found Peter standing there. He had just quit his job as CEO of Liberty Media, he told me. He'd recently turned the same age as his father when his father had suddenly died of heart disease. Peter had a sense of foreboding that he, too, would soon die. Just forty-seven years at the time, he had decided to rent a train car and somehow use it to explore the United States with his wife and three young children. Once he came back from that journey, he started up his own investment company. I discussed some of his projects with him, but I could tell his heart was not really in it. Then, at perhaps forty-eight, he was diagnosed with stomach cancer.

Peter died at fifty-one. The last time I saw him was at the 1999 Sun Valley Conference, where he made an appearance, already gaunt from dealing with the cancer. Somehow, in 2002, I learned he was in the hospital and had just a few days left. I sent him an email, just hoping he might see it. As it turned out, he had his BlackBerry with him. He not only read it but wrote back, just a few minutes later: "I will see you on the other side." I cried. Just a few hours later, Peter died, much too young. His death reminded me of the preciousness of life. You never really know just how much time you have left. I decided then that I would not put off what was important to me, just as Peter had, not because he knew he would die young but because he knew he could die young.

When I left Intel in April 1999, I asked Herbert Allen if he would let me attend the next conference, joking that I was about to be "delisted" because I knew that I had been invited based on my position at Intel. Herbert agreed. All the companies were listed alphabetically. I listed my company as The Avram Miller Company, which was placed in the conference materials just above The Walt Disney Company. I really enjoyed that.

As is well known, lots of business deals actually do get done at Sun Valley. I introduced Andy Grove to Steve Case of AOL at Sun Valley. Steve had asked me if Intel would join AOL in acquiring UUNet, a nonprofit organization that provided the US Internet backbone and was about to become a for-profit business. Steve also wanted Intel to acquire his dial-up network as part of the deal. Intel would have ended up owning a large share of AOL. I hoped that meeting Steve Case in person would lead Andy to support this concept, but it didn't. Instead,

Andy decided to place a bet on AT&T. This was just one of many times I failed to get Intel to support the consumer Internet.

One year, Renée James, Andy Grove's technical assistant and later Intel's president and COO, set up a computer lab where she gave tutorials on how to use PCs. The demos were primarily attended by the spouses of the various CEOs at the conference. It was pretty effective in introducing the PC's potential to this influential group.

The conference continues today, attended by people like Mark Zuckerberg. Herb Allen's son, Herb Allen III, runs things now.

PART III

NEW HORIZONS (1992–2002)

The development of residential broadband and the creation of a new medium.

11. THE PC IS IT

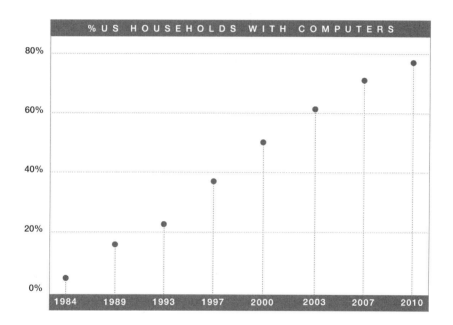

Personal computers become standard in the home.

The last thirty years have seen dramatic changes in how consumers interact with technology. The development of high-speed residential broadband, powerful computers, and other intelligent devices in the home, coupled with the Internet, have transformed how we work, play, and communicate. While others will have more detailed knowledge of specific aspects, I may have the best overall understanding of the totality, the big picture. My participation in the development of residential broadband was the most important and exciting activity of my career. I am extremely fortunate to have been a part of such a transformational development.

My involvement with the development of residential broadband primarily spanned the 1990s, which will be my focus here. But I do want to put the past thirty years in context. While this transformation took place all over the world, I will mostly touch on the events in the United States.

We can divide up these years by decade: foundation, transformation, and ubiquity.

PHASE I: FOUNDATION (1990–2000)

From 1990 to 1999, the share of US homes with personal computers spiked from 15 percent to 50 percent. Penetration of dial-up online services went from next to nothing to more than 30 percent of homes. In the process, one of the early online services, AOL, became worth more in terms of market capitalization than any media company, more than Time Warner and ViacomCBS combined, more than all the leading cable companies combined, and more than all the American telephone companies combined.

Back in the 1990s, television was still analog, and most people used a landline and not mobile telephones to make phone calls. In 1990, 57 percent of homes received television via cable. By the end of that decade, cable penetration was about 70 percent. As the '90s began, cable companies were yet to realize that home computers would transform their business. Phone companies did not understand how both home computers and mobile devices would make their business almost unrecognizable. It was a period of rapid change with winners and losers.

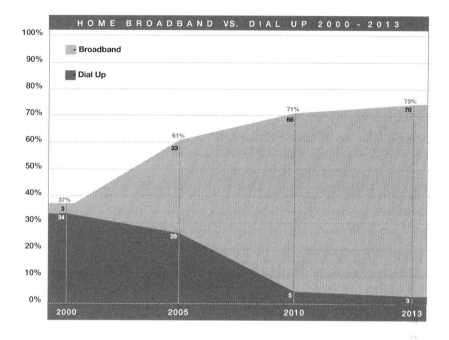

During phase I, both cable and telephone companies built out the infrastructure for the residential high-speed broadband Internet access many Americans now take for granted. The cable companies initially built this infrastructure for the digital transmission of television, including high-definition TV (HDTV). Cable companies invested more than $100 billion to upgrade their networks. At the same time, technology, media, and communication companies spent billions in a misguided and largely futile effort to bring interactivity to the TV.

In the 1960s, the US Department of Defense developed network technology to connect computers, called ARPANET, used at first by the military and then by educational and research organizations around the world to share information. By the 1980s, ArpaNet evolved into what we today call the Internet. Only in 1993, with regulatory changes by the Clinton/Gore administration, was the Internet opened to commercial use. The first advertisement on the Internet appeared on October 24, 1994. Advertising revenue would be extremely important to the development of the consumer Internet.

> Advertising is going to be the killer application for the
> interactive PC . . . the thing that's going to pay for all
> of this.

—Avram Miller, quoted in *Advertising Age*, June 27, 1994

Nasdaq index 1990–2000

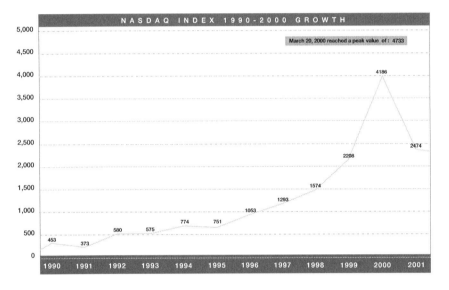

The NASDAQ index grows by 4,000 percent in ten years.

The consumer-based Internet we now know, which most of us use daily, began with the founding of Netscape and Amazon. Portals like Yahoo, Lycos, Excite, and MSN fought for consumers and the advertising dollars they brought. Valuations for Internet companies created a new gold rush, centered in Silicon Valley. Companies like Intel, Microsoft, and Cisco reached valuations that dwarfed traditional industrial companies.

I started working on the development of residential broadband in 1992 and continued until I left Intel in 1999. I led Intel's efforts to pull together all the elements that were required to build the foundation on which a new medium would be born. While Intel was already playing a major role in the development of all aspects of the personal computer industry, this would be the first time we would venture into the development of the network that would support the growth of home computing.

By the end of the 1990s, television was increasingly transmitted digitally, with high-definition TV (HDTV) starting to roll out. Digital video recorders such as TiVo would make possible the easy, automatic time-shifting of television programs. The cable industry, which had significantly upgraded its network to provide for such services as on-demand HDTV, was now in a position to offer cable customers high-speed Internet access for their personal computers. Telephone companies could offer high-speed ADSL modems to more than half of their customers.

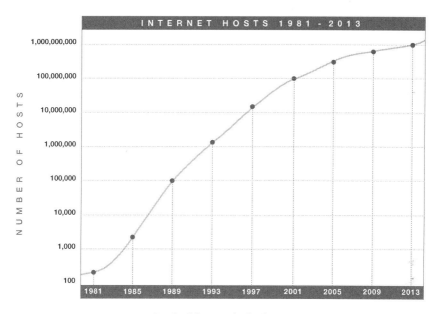

Graph of the growth of websites

PHASE II: TRANSFORMATION (2000–2010)

High-speed broadband Internet access grew from being in about a million US homes in 2000 to more than 55 million by 2010. The number of devices within homes also grew. Wi-Fi was introduced, making it easy to connect multiple computers and other devices without having to run cables within homes. Consumers were discovering that it was not enough to have one computer per household.

The speed of the connection enabled more complex web pages, and advertising became the dominant source of income for consumer Internet sites. E-commerce growth began to significantly impact shopping, with Amazon becoming a major force in the retail sector. The Internet became the platform of choice for online communication, and the number of websites grew dramatically. As a result, search became the most critical application, reducing the importance of portals and leading to Google's commercial ascendance. The second half of phase II saw the Internet become a media distribution platform for both audio and video.

— we went from many people sharing one computer to many computers sharing one person

PHASE III: UBIQUITY (2010–2020)

The transition to this period actually began with the 2007 release of the iPhone, which invented the smartphone category of device that supported Internet browsing by both cellular data and Wi-Fi connections. Smartphones led to substantial growth in mobile computing, leading the way for consumers to be always connected: everything, everywhere, all the time. Smartphones also tremendously impacted the growth of the Internet in developing countries. Mobility would be a critical contributor to the growth of social media applications, as well, such as Facebook.

A REVOLUTION IN RESIDENTIAL INTERACTIVITY

Three intersecting technologies—semiconductors, fiber optics, and software—combined to provide the foundation for massive changes in technology starting in 1990. The first two are made from sand. The third, imagination. This major shift in residential technology had profound implications for how people in the twenty-first century interact with their world, surely touching nearly every one of you directly.

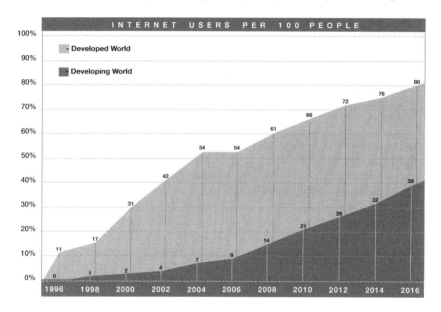

When I was growing up in the 1950s, the only electronic connections homes had to the outside world were radio and television broadcasts and the telephone. Our news, just like our mail and possibly even our milk, was delivered every morning to our home. There were just three television networks: NBC, ABC, and CBS.

Almost every house in 1950 had a radio, just about 60 percent of homes had a telephone, and most had a record player. Radios were large pieces of furniture, though soon the transistor radio would provide the first mobile consumer communications device.

Fewer than 10 percent of American homes had televisions in 1950; by 1960, more than 90 percent did. Another thirty years would elapse before the introduction of the next major home electronics device, the videocassette recorder (VCR), and that just extended the television in the same way as the answering machine extended the telephone. (In case you are especially young, both recorded signals for later playback that would otherwise be lost immediately after transmission.) Video game consoles, which came to prominence in the 1980s, also augmented television. Games consoles were the only interactive devices, and they were used primarily by young men.

Electronic calculators did not exist in 1950. The world's few computers occupied large, air-conditioned rooms and were serviced by specialized technicians who usually wore white coats.

Forty years later, in 1990, not that much had changed with respect to home communication. Homes had telephones and televisions. Telephones had buttons instead of a rotary dial. You could now call someone almost anywhere in the world, sometimes even while you were driving your car. Television was broadcast in color, and there were many more channels. In the United States, most homes, 53 million, were served by the cable television industry, which had begun not only retransmitting broadcast television but also creating new television networks, such as CNN, HBO, and ESPN, that could only be received by cable or satellite.

While we had first sent men to the moon twenty years earlier, it still took a week to mail a letter from San Francisco to Paris. That would all change with the introduction of email.

By the end of the 1990s, more than half of US homes would have computers, and most of these homes also had cable television, as would later prove important. In offices, businesses were rapidly deploying networked personal computers. Increasingly introduced to the capabilities of personal computers at work, people began to want the same capabilities in their homes.

At first, most thought that interactive media would come into the home through the television set, since almost every home had a least one television and few had personal computers. Cable, telephone, computer, and software companies spent billions in the pursuit of this flawed idea. Many executives, including Microsoft's Bill Gates, felt that the most important real estate in the home is the square foot on top of the TV set.

At Intel, we first experimented with interactive television in 1992 ourselves, primarily to keep Microsoft happy. But it became clear to us that the PC and not the TV would become the center of home inter-activity. Of all the major technology and media companies, Intel was alone in thinking this, even though it seems so obvious now. At that time, the resolution of television sets could not support the kinds of applications consumers wanted. People went to their living rooms to relax, looking for a passive interaction: a "ten-foot" experience. The hardware required to enable interactive television would also be too expensive to ever be appealing. But all of this only became clear after billions of dollars of wasted investment by companies like Time Warner and Microsoft.

Led by Andy Grove, we adopted the mantra, "The PC is It," mean-ing that the PC would be the dominant interactive device in the home. This represented a major shift in thinking for Andy; in 1991, he had insisted that there was no consumer market for personal computers. He ridiculed me in front of others for holding this view. But when the Pentium processor was introduced just two years later, with consum-ers its initial target market, it was clear that the winds had changed. To his credit, Andy shifted and put his full force behind making the personal computer the dominant interactive device in the home. We actually became a kind of team in advocating this position.

As I write this book in 2021, the PC remains the dominant interac-tive device in the home. The smartphone is simply an evolution of the PC, even if it's challenged the primary position of the PC for younger people and in the developing world. Interactive television was envi-sioned to involve such things as shopping, reading newspapers, and communicating with friends. This has never happened. Today, the primary interactive activities related to the television set are selecting which programs to watch and playing video games, with devices like the Apple TV and Google Chromecast just glorified remote controls.

> For Avram Miller, vice-president of corporate busi-ness development for Intel, the big story of the present—personal computers—will remain so, far into the future. "The personal computer," he says, "will be to the 21st century what the automobile was to the 20th.

It will reorganize the way we live, play, and spend our lives. How it will affect a particular business will vary with the business. But one thing appears obvious— some of the hottest places for doing business will not be places at all, but will exist in cyberspace. There will be a whole new world of virtual products and services that live only in this other world, which we will create as we go along . . . I know it sounds wild. It's as hard for us to understand as the airplane was in the 19th century. Yet now we fly on planes every day. Cyberspace will be a world 21st-century man will feel at home in."[1]

YOU'VE GOT MAIL: AMERICA ON HOLD

The PC's success in the home required connectivity. Only a limited number of productivity applications could stand alone for home use, like word processing, home accounting, and tax preparation. Consumers began to dial into online services—America Online (AOL), Prodigy, and CompuServe—primarily to communicate with each other either by sending email or in online chat rooms. Although some forms of instant messaging existed on time-sharing systems, the Israeli product ICQ would introduce instant messaging to the consumer world. AOL paid $407 million in 1998 to acquire the company behind ICQ and its product. Later, it replaced ICQ with AIM, which was the leading instant-messaging system. Microsoft brought out Microsoft Messenger in 1999 to compete with AIM. Instant messaging remains a very important function and is one of the most important capabilities of smartphones, and messaging apps have about 3 billion active users.

Of course, AOL could never have been developed without the existing combination of personal computers and dial-up modems. The computer industry had essentially hacked the telephone industry to develop modems in the '80s. Some smart engineers figured out that they could use tones in the same range of frequency as speech to

---

1.   Michael S Malone, "Chips Triumphant," *Forbes ASAP*, February 26, 1996, p. 74.

connect computers. Something similar had been done earlier to create fax machines. In a sense, then, engineers could fool the telephone network to carry a form of communication not previously imagined. The fact that computers could communicate from one location to any other location around the world, provided they both had telephone service, was a miracle for the computer industry. Without a way to use the telephone system, originally designed for voice communication, the development of the home computer market and especially online services would have taken many more years to emerge.

The technology that enabled such communication between computers, the modem, was named after the way it "modulated" the frequency of the signal (and, on the other end, "demodulated" the same signal) within the range of the human voice in order to carry information. This was very clever. The telegraph, which preceded the telephone, was baseband; the signal either existed or did not, and information was conveyed by a pattern of signals in time (known as Morse code).

But the telephone was broadband, modulated, meaning a signal could shift in frequency! Companies like Hayes and US Robotics developed significant businesses supplying modems. The first modems had acoustic couplers, because the US telephone monopoly AT&T wouldn't allow non-AT&T equipment to connect to their infrastructure. The official AT&T handset would thus be placed into a cradle on the modem. Later, when AT&T was forced by regulation to allow it, modems were connected directly to the telephone system using the same kind of cable used for telephone equipment. In some places, though, the phone might be wired directly and could not easily be removed or replaced. I used to travel throughout the world with tools so I could connect my portable computer's built-in modem to whatever phone system I might encounter in various hotels.

Telephone modems were typically 2,400 baud (bits per second) during this period. Some fortunate people might have 9,600-baud modems. To understand just how slow this was, consider that it would have taken hours, if not days, to download just one average smartphone photo. Eventually, but only by 1999, modems reached 56 kbits, the maximum possible using the range of frequencies in the human voice. They were too expensive for most consumers, and there was no standard.

Modems were connected to PCs through serial ports, unless the PC had one built in. The IBM PC had a serial port, and the Apple II had an optional one. Many other early home computers, like the Commodore PET, also had serial ports. These ports were not very fast, but they did not have to be, as modem speeds were limited by the frequency range of the telephone network that was designed for voice communication. While some computers also had parallel ports for printers, these were not much faster than the serial ports and, in many cases, operated only in one direction. The lack of a high-speed connection became a major obstacle to the creation of high-speed residential networks.

Connecting to an online service, like AOL, required the dedicated use of a telephone line. Most people only had one telephone line, so that line would not be available for voice calls when they were connected to the online service. Homes that used online services a lot often installed a second telephone line, which added cost. Trying to connect to an online service could be a lesson in frustration. First, you would have to turn the computer on. Several minutes would pass before you got to the point where you could ask the computer to connect. Then, the computer's modem would dial a telephone number for the online service, which was often busy. The computer would try again and again, dialing different numbers, sometimes dozens of them.

At first, AOL charged customers $5 per hour. In 1996, they went to a flat monthly rate of $19.95, or more than $30, adjusted for inflation. With a flat charge, customers began to utilize the system much more. To cover the additional resulting costs, especially connection charges and equipment, AOL began to sell advertising (banner ads).

AOL, it turned out, became the dominant consumer online service. Unlike so many Internet companies in the late 1990s, it had real revenues, reporting $4.8 billion revenue and 20 million paying customers in 1999 without a single dollar of debt. In 1998, AOL acquired Netscape for $4.2 billion in stock. Lucky Netscape shareholders who held on to the AOL stock they received saw the value of those shares increase fourfold before the Internet bubble burst in 2000.

By 1999, AOL had 18 million users. Steve Case, AOL's CEO, knew that the valuations of AOL and all the other public tech companies were crazy. Desperate to buy some real assets with his inflated stock, AOL eventually merged with Time Warner. While generally considered one of the greatest business disasters in history, as it certainly was for Time Warner shareholders, AOL shareholders, including Case, were able to effectively and profitably convert their overvalued AOL shares into Time Warner shares. Still, the merger was dysfunctional, and hastened the decline of both companies. Time Warner ultimately divested its AOL business in 2010; in 2016, Verizon acquired AOL. AT&T would ultimately acquire Time Warner in 2018.

— when your stock is overpriced,
use it to buy real assets

Netscape was the other company that played a leading role in creating the consumer interactive market, which is why AOL bought them. Netscape enabled companies like Amazon, Yahoo, Lycos, Excite, and many more to reach consumers. Once Microsoft added Internet Explorer, all home PCs gained access to the consumer web. But AOL and Netscape, not Microsoft, had ignited the consumer market.

AOL's dominance was toppled by the impact of broadband, the wider Internet, and, most importantly, Google search. In a concept known as a "walled garden," early consumer online services, including AOL and Prodigy, only allowed their own applications and connections. You could just chat with or email someone who was also on AOL. I worked with both of these companies to open up their systems to the Internet, but the release of the Netscape web browser really drove the adoption of the consumer web over the early walled-garden online services. Meanwhile, web portals such as Yahoo and Excite, and later AOL, curated web content and provided important functions, such as email. Google's powerful search undermined the need for curation, since their PageRank algorithm could determine which sites were the most popular and important and show these first (in the process, of course, making them even more popular).

MICROSOFT WAKES UP (SORT OF)

Microsoft was late to the Internet party, perhaps because Bill Gates was busy chasing interactive TV. Once he realized that the web browser platform could undermine the Windows operating system, he wrote a memo to his executives requiring all of Microsoft to refocus on the Internet. Microsoft licensed a browser from a company called SpyGlass, which, in turn, had licensed the original Mosaic browser from the University of Illinois, written by Netscape co-founder Marc Andreessen. This became Internet Explorer 1.0, which Microsoft then bundled with its operating system. Netscape, which had a better product, required consumers to download their browser, which was difficult to do using dial-up modems. Microsoft began to systematically crush Netscape. This became the subject of various legal and anti-trust actions taken by the federal government. Eventually, the United States required Microsoft to bundle both Explorer and Netscape; by this time, however, it was too late for Netscape, which AOL eventually bought. Netscape did not actually play an important direct role in creating the consumer Internet. Rather, it became primarily focused on the enterprise market. Microsoft bundling their Internet Explorer with Windows had a very powerful effect. By 2003, Microsoft had a 95 percent share of the browser market.

As for AOL, Microsoft at first wanted to acquire them. In negotiations at Microsoft's Redmond headquarters, the two sides couldn't come to terms, mostly because Bill couldn't get along with the AOL people. After ending negotiations, Bill instructed Microsoft to build a competitive system, called the Microsoft Network or, more commonly, MSN. AOL successfully defended itself from attack by MSN largely because Microsoft remained focused on defeating Netscape, which it (correctly) saw as a greater strategic competitor. Eventually, in fact, AOL and Microsoft made a deal: Microsoft would bundle AOL's software with Windows, and AOL would bundle Internet Explorer.

Microsoft, unlike Intel, did not limit its thinking to just the desktop. They chased some dead ends along the way, though, like WebTV. In 1995, Steve Perlman, someone I knew pretty well, had started WebTV to build a product for browsing the Internet on a conventional TV. Though I looked at WebTV for Intel, we did not invest. In fact, I convinced both Rupert Murdoch and (separately) Richard Li of PCCW not to invest. Microsoft acquired WebTV in April 1997 for more than $500 million. The product was renamed MSN TV and actually continued in operation until 2013 but had no real impact.

Others offered similar products. Liberate, originally set up as a division of Oracle in 1995 and spun out in 1996, developed set-top box technology to license to others. AOL licensed their technology for AOL TV in order to compete with WebTV. None of these or other TV-based web set tops were successful. The low resolution of the television itself, the poor communications technology that then existed, and the growth of the PC market all conspired to reduce the appeal of this type of device. Now, even in 2021, when TVs themselves are "smart" and have broadband access, the amount of time spent browsing the Internet from the TV is negligible, as is the number of people who ever use a television for this purpose.

> No matter how much computing power you put on top of the TV, you cannot put more pixels on the TV's screen.
>
> —Avram Miller, *Computer Chronicles*, 1995

12. THE BATTLE FOR EYEBALLS AND EARS

"These are going to be the range wars of the 21st century," says Avram Miller, vice president for corporate business development at Intel Corp., "and not everyone is going to survive."

—*Newsweek*, May 30, 1993

RANGE WARS OF THE TWENTY-FIRST CENTURY

While we at Intel worked to make the PC the primary interactive device in the home in the early 1990s, the cable and telephone companies were each trying to get into the other's business. Remarkably, neither industry paid much attention to the growing number of personal computers being sold for home use, nor did they consider the growth in online services to be the threat it actually was. Every telephone company executive had a television. Every cable company executive had a telephone. Almost none had a PC at home. Most executives did not even operate their own PCs but rather had their assistants print out emails.

The cable companies' primary strategy involving digital technology was to use video compression to provide more television channels over cable and create new video-on-demand services. Their primary competitors, as they saw it, were satellite television companies, such as DirecTV and EchoStar. Utilizing digital compression technology, satellite television could offer many more channels than cable companies could using analog technology. Though the cable companies had already begun moving to digital transmission, this would only put them on par with satellite. Cable executives thought that television-based interactivity could create not only new sources of revenue but also a competitive barrier to entry, since it was not easy to send a signal back through the satellite as the nature of satellite communications creates long latency.

Additional competition from the cable companies' perspective included videocassette and digital video disk (DVD) rental companies, such as Blockbuster. They certainly could never have imagined that one day a company called Netflix, which started by sending DVDs through the mail, would have more subscribers than all the cable companies combined.

Presenting another competitive threat to the cable companies—and one with access to almost every US home—the local telephone companies (each a local exchange carrier, or LEC) desperately wanted to find a way to provide television service over their telephone lines. Local telephone companies had large, continuous franchises across multiple states. While cable television companies, like the telephone companies, enjoyed monopolies in the areas they served, cable access rights were granted by local municipalities. Since cable companies had to either dig up streets to put in cable or put their cable on top of telephone poles (actually owned by the electric utility companies or sometimes the local phone company), cities did not want to have more than one provider of cable television. Cable companies would compete with each other to get the franchise for particular areas, which led to quilted ownership by geography.

Cable companies were termed MSOs (multiple-system operators) for this reason.

Cable companies even worried that the electric utilities would find some way to use their connections to the home to provide video service (and there were some later attempts to do this). Some even thought that water utilities could become competitors. It would be quite some time before wireless technology would worry any of these monopolistic utilities, which were not used to having competition.

While the local phone companies lusted after the cable companies' television business, the cable companies coveted the local phone business. Households in 1990 spent almost ten times more for their telephone service than their cable connections, a per-customer revenue disparity that presented the cable industry with a huge growth opportunity. The cable industry began to work with AT&T, the major long-distance carrier; Sprint; and others to develop a way to use their cable plant to provide telephony services. Their approach to telephony over cable was initially very specific; industry executives believed that television, telephony, and the Internet were three distinct uses of their network that required three unique technical solutions. By contrast, I understood that television and telephony were, in a sense, just network applications that could use Internet protocols. Eventually, the cable industry came to this conclusion regarding voice, announcing a standard called PacketCable in 1997 for voice. The industry still has not fully integrated video. Though hard to imagine now, for most of the 1990s, the cable industry prioritized telephony over providing high-speed Internet access. There were a few exceptions, including Mark Coblitz and Steve Craddock at Comcast, who realized that, one day, broadband would surpass the revenue of television service.

*— cable plants were like oil fields before
the invention of the automobile*

The desire to provide local telephone capabilities also motivated investments by AT&T in cable companies, including its acquisitions of TCI in 1997 and MediaOne in 2000. AT&T, which was the original parent company of most local telephone companies before its 1982 antitrust breakup, wanted to return to its roots and use the cable infrastructure to provide local telephone service.

Of all the brain-dead companies in the 1990s, none, in my view, were as backward-looking as AT&T. Before its breakup, AT&T was a highly regulated and successful monopoly. AT&T wanted to enter the computer business, which they felt would be a more attractive industry, but the US government required that AT&T first divest themselves of their local telephone business. This created eight local phone companies, known as Regional Bell Operating Companies (RBOCs). In its remaining long-distance telephone business, AT&T faced competition with companies such as MCI.

Telephony required two-way connections on the cable plant, and the lack of quality upstream bandwidth on cable also presented a major barrier to the deployment of residential broadband connections to PCs in homes. Telephony, the desire for more TV channels, video on demand, interactivity, and ultimately HDTV all encouraged major investment (more than $100 billion) by cable companies to upgrade their physical plant to utilize a combination of coaxial cable and fiber optics. They did not at first realize that they were actually building out a broadband network for home computers, a critical component of what was often called the information superhighway.

Meanwhile, though the local telephone companies would fail at their video service efforts, we used the resulting technology to develop DSL (digital subscriber line) and put the local phone companies, too, into the residential broadband business. The local phone companies had thought that they could use their market presence with consumers to offer a video alternative to cable, and they also believed they could provide interactivity. But they did not actually have enough bandwidth for quality video; worse, they could not guarantee quality of service, because the amount of bandwidth they could deliver depended on factors such as the distance of the home from the central office.

— *landline: a race to the past*

In 1995, Americast, BellSouth, SBC Communications, GTE, and The Walt Disney Company formed a consortium to offer television services over copper telephone wire. They placed a billion-dollar order with Zenith for set-top boxes. The venture started winding down service in 1996. It was too expensive. By 2000, it had totally closed down. Another consortium, called Tele-TV, comprised Bell Atlantic, Pacific Telesis, NYNEX, and Creative Artists Agency. Started in 1995, it closed down in 1997. Bell Atlantic, for one, had put in a large order for set-top boxes.

I still find it hard to understand why the local phone companies thought they could compete with cable companies for video delivery using their copper telephone networks. The cable companies had an inherent technical advantage with their coax versus the phone company's copper twisted pair. This meant the phone companies were trying to race a jet airplane on roller skates. It was a strange time, this race to the past, with the telephone companies wanting to get into the TV distribution business and the cable companies wanting to get into the telephone business. Neither had any idea that the future would be dictated by high-speed residential access to the Internet and, later, mobile wireless devices. Sadly, my own company, Intel, failed to exploit this latter opportunity. At that time, cable companies were dependent on their vendors such as General Instruments and Scientific Atlanta for engineering. It was not until the early 2000s that cable companies would develop their own engineering organizations.

Alongside this story, two young semiconductor companies just fifty miles apart in Southern California were laying down the technical foundation of the revolution to come. One, Broadcom, developed the chips that both made cable modems and digital television possible. The other, Qualcomm, developed the technology underlying mobile computing devices, such as the iPhone. Intel could have been either of these companies. Intel was an early investor in Broadcom but passed on an investment in Qualcomm after I had met with them. We only invested in Broadcom because they made chips for Ethernet. I was very impressed with Qualcomm when Les and I met with its founders in 1991. Les did not believe that Intel would be interested in Qualcomm, so I asked him if I could invest personally in their IPO. Happily, he agreed, because Qualcomm was not considered strategic or a competitor. Today, Qualcomm sells many of the microprocessors (actually systems on chips) for mobile devices, and is expanding into server technology, taking market share directly from Intel.

THE FIVE-HUNDRED-CHANNEL UNIVERSE

Even though the phone companies' DSL was an inferior technology compared to cable modems, I knew it could be deployed earlier and in areas where cable was not present. In particular, I thought it would be the dominant technology in countries like Germany where cable television had not really developed. Most of all, I aimed to create significant competition between the cable and telephone industries with respect to residential broadband.

Both the cable and the telephone companies were determined to provide what cable executive John Malone termed the "five-hundred-channel universe," increasing the footprint of what FCC's then chairman Newton Minow had called "a vast wasteland" in 1961, a nickname that entered general usage to describe American television. I had to convince the cable companies that they were not in the television but the communications business, even as I convinced the telephone companies that they, too, belonged in the communications business and not the voice business. The obstacles to overcome were many. I spent several years meeting with CEOs of both cable and telecommunications companies, trying to inspire them to move into the twenty-first century. I traveled constantly, not only in the United States but throughout the world. I spoke at both cable and telephone company conferences, as well as at conferences organized by the major investment banks.

I learned a great deal about the cable industry from colleagues at General Instruments (GI) as part of a collaboration between GI and Intel to develop a smart set-top box, which I describe in Chapter 9. I needed to understand the telephone companies just as well. I hired a consultant to teach me about that industry, including the key technologies. I even visited telephone company facilities to understand their equipment. I wandered around exhibitions, talking with equipment manufacturers. And, most of all, the Intel Architecture Labs gave me a great deal of support. Brilliant people there like Kevin Khan and Ali Sarabi patiently helped me to understand some of the complex technical issues involved.

The cable companies were willing to take more risks than were the phone companies. Once they understood the potential to provide broadband, they strongly embraced it, although not as quickly as I wanted. Unfortunately, the phone companies moved more slowly than I had hoped and did not really push the cable companies to move much faster.

> Their strategy for tomorrow is to make today last.
> —Avram Miller, on the telephone companies,
> *Forbes*, February 13, 1998

The computer industry wanted to supply the arms for the expected war between the cable and telephone companies. Microsoft, Oracle, Sun, Silicon Graphics, Hewlett-Packard, Compaq, Sony, and many others had programs to address the interactive TV opportunity, as did communication equipment suppliers like Cisco and Motorola. Semiconductor companies, like Broadcom, also focused on this opportunity. At Intel, our objective was different. We just wanted the PC to be the dominant interactive device in the home.

BILL GATES CALLS ANDY GROVE, AND I AM ON MY WAY TO REDMOND

My involvement with broadband had a rather strange beginning. In 1992, I was sitting in my cube (cubicle) on the sixth floor of the Robert Noyce Building (RNB) in Santa Clara, named after Intel co-founder Bob Noyce, who had died two years prior.

While Intel always claims it has no headquarters building, RNB came pretty close, just as the Bowers Avenue building had before. Though the sixth floor was the top floor of the building, Andy Grove, the CEO, and Craig Barrett, the COO, for some reason had offices on the fifth floor. But even they did not really have offices. RNB had no actual offices, nor did almost all Intel facilities. Intel prides itself on its egalitarian rows of cubicles. When I first joined in 1984, I found that strange, since I'd had my own office since 1969. Now, if I stood up, everyone could see me. And, of course, everyone could hear everyone else's conversations.

Given my relatively senior position at Intel, my cube was large enough not only for my desk but also for a table, where I could meet with one, maybe even two, other people. I also had a window to the outside. Now that was real status! Next to my cubicle was a similar one belonging to my boss, Les Vadasz.

On that fateful day, Les walked into my cube. He had just spoken with Andy, and Les explained that Andy wanted my help on a project. I should go see Andy, Les said. (Andy was following Intel protocol by getting Les to agree in advance that I could work on Andy's project.)

Andy's cube was just a bit larger than mine, though he also had his own conference room. As we spoke in his cube, he explained that Bill Gates had called him that morning. Microsoft was very interested in developing products for the consumer market, Bill had said, but Microsoft had failed to find anyone at Intel with similar interests. Since Intel apparently had no desire to participate in the consumer market, Bill told Andy he assumed that Intel would not mind if Microsoft worked with AMD (Intel's primary competitor) to develop consumer products. Bill, of course, said this in order to get Intel's involvement. He knew how Andy felt about AMD. "What do you mean, no one at Intel is interested in the consumer space?" Andy replied. "One of our VPs, Avram Miller, is working full-time in the consumer space!" And so I found myself in Andy's cube.

Andy had reluctantly agreed to let me work on the consumer market. He thought it was a waste of time; consumers would not be interested, he was convinced at first, in using computers at home. Andy had a history of being slow to understand the potential of the PC. In 1991, when Kirby Dyess began to introduce personal computers to Intel's offices, Andy had the same reaction. At that time, Intel ran on mainframes and minicomputers. Andy saw no need for personal computers in the enterprise. Later, of course, even if a few years late to the party, Andy would become the world's biggest advocate for personal computers.

Bill Gates knew me already. We had first met in 1981, when he visited Digital Equipment Corporation at the invitation of Barry Folsom, the developer of the Rainbow, which ran MS-DOS. A year later, in 1982, I met Bill again at the PC Forum. At the time, I was still at Digital.

Bill asked Andy to send me up to Microsoft's headquarters in Redmond, Washington, as soon as possible to meet with his consumer product point person, Rob Glaser. Andy had agreed, he told me. He then clearly explained that my primary mission was to keep Microsoft away from AMD. Though Intel's fortunes were by this point hitched to the PC, he emphasized that he still thought additional attention to the consumer market would be a massive waste of time. So, off I went to Redmond with the mission of keeping Microsoft away from AMD. I still thought I might have found the breakthrough I had been looking for: finally, a chance to work on the development of consumer products at Intel! I was not going to miss this opportunity.

That was my first time visiting the Microsoft campus, which was very different from Intel facilities. Many people had actual offices, with doors and windows. Lots of food and snacks were available. Tim Yiu, who worked for me, accompanied me up to Redmond, where we were met by Rob Sullivan, Intel's account executive for Microsoft. We first met with Rob Glaser and members of his staff. A one-on-one meeting followed between Glaser and me. He was just thirty years old at the time and had joined Microsoft nine years earlier, at the age of twenty-one. He was already very wealthy and clearly very smart. I was forty-seven years old and beginning to learn that I was no longer the youngest person in the room. (In fact, everyone at this meeting was younger than me.)

Rob proposed three projects for Microsoft to execute in some kind of partnership with Intel. The first was interactive TV. The second was a device to compete with the Apple Newton and other personal digital assistants, which we ended up calling WinPad. The third was a game machine that would use CD-ROMs. I agreed to have Intel work on the first two opportunities but declined to do the game machine. Many years later, Microsoft would introduce the Xbox.

While Rob Glaser is a very intense, very smart man, he was not always an easy person to work with. We still exchange messages from time to time. Six months after we met, Rob left Microsoft after a major political battle with Nathan Myhrvold, who at that time led Microsoft Research. Nathan would eventually become Microsoft's first chief technology officer (CTO). Craig Mundie, who had joined Microsoft about this time and reported to Myhrvold, was then assigned to work with me on consumer opportunities. Craig eventually also became Microsoft's CTO. Craig had no consumer background at all; his experience was in high-end minicomputers. I don't believe that any of Microsoft's television-related products were ever successful until the gaming device, the XBox, was introduced in 2001.

INTEL/MICROSOFT JOINT EXECUTIVE SESSIONS AND SCREAMING MATCHES

Throughout the 1990s, Microsoft and Intel held joint executive meetings every quarter to discuss areas of cooperation or conflict. The location would alternate between Redmond and Santa Clara. The key participants were, of course, Andy and Bill; the other participants varied. Ron Whittier often attended; at that time, he led the Intel Architecture Labs. Steve Ballmer was often there. Probably at Bill's bidding, upper Microsoft and Intel management decreed that Rob and I should attend these meetings to report on our progress. Then, Rob was replaced by Craig Mundie. Bill was strongly committed to the consumer space. Andy could not help but show his disinterest. It was an amazing opportunity for me to attend these meetings with the top management of Intel and Microsoft. The sessions were often very contentious, to say the least.

Bill Gates wanted to absolutely dominate all forms of interactive computing: not only the desktop, but the TV top (back then, TVs had tops) and the attaché case (PDAs were too big to fit into your pocket). Microsoft totally failed at both—but not for lack of trying. Intel, on the other hand, only thought that the PC would be successful. Intel was right for a very long time and then totally missed the move to mobile computing, including turning down Apple's request to manufacture the processor for the iPhone.

Gates did not want any other company to gain even a toehold as a software platform. In my entire career, I never met anyone as competitive or aggressive as Bill was then. He would literally stand so close to you that his nose would almost touch yours. He would scream at people. He screamed at me more than once. I learned not to back away. Though I am not naturally an angry or aggressive person, I sometimes raised my voice to him. Once, he screamed at me in front of the joint group of executives. He had called me the day before, and my assistant told him I was not available. "Don't you have a fucking cell phone?" he demanded. I calmly replied, "I do, Bill, but yesterday was Yom Kippur. On that day, the only call I'd take would be from God. And you are not God." In social situations, frankly, he was a nicer person. Fortunately, the Bill Gates we all see as the leader of the Gates Foundation appears to be a very different person than the Bill Gates I knew when he worked at Microsoft.

FULL SERVICE NETWORK: A PIPE DREAM

Microsoft was just one of many companies interested in interactive television. Warner Communications, which later merged with Time Inc. to become Time Warner, had a history of experimenting with interactive TV. Warner Cable's QUBE, launched in Columbus, Ohio, in 1977, provided some basic interactivity and video on demand. This capability helped Warner win cable franchises, and QUBE was brought to some other communities. While it might have grown into something significant, a combination of consumer privacy issues and serious financial losses by Warner Communications overall led to QUBE's termination in 1984.

But Time Warner retained its desire to provide interactive service, developing a concept they called the Full Service Network (FSN), a TV-focused interactive system. With the strong support of CEO Gerald Levin, Time Warner put out a request for proposal (RFP) to many companies, including the leading computer companies. I remember reviewing the FSN RFP and deciding there was nothing really interesting in it for Intel. While we did not respond, the RFP certainly got a lot of attention.

As a media company with a cable system, Time Warner wanted to become the primary supplier of interactive television for the entire cable industry. The FSN was their vehicle for achieving this objective.

> Ed McCracken, Silicon Graphics' boss, reckons that the PC has no future in interactive multimedia: "I believe that fewer than 15% of multimedia users will interact with the information superhighway via their computers; the other 85% will use the television. And the capabilities we're building into TVs will make them more powerful than any PC."
> — *The Economist*, September 17, 1994

As technical partner, Time Warner selected Silicon Graphics, Inc. (SGI), a company making high-end computer workstations. (SGI had been founded by Jim Clark, who would go on to found Netscape.) SGI workstations were commonplace in the film and television industry, so they had an existing connection to Time Warner. FSN was the kiss of death for Silicon Graphics. The FSN project was terminated in 1997, in part because SGI's set-top boxes cost more than $2,000 in today's dollars. Ed McCracken, the CEO, resigned around the same time. While the demise of any company has many causes, the time SGI wasted on the FSN at a pivotal moment for the computer industry could not have helped. SGI then entered a long decline, ending in its 2008 bankruptcy.

PANDORA'S BOX

Gates was very focused on what he considered the most valuable real estate in the home: the top of the television. I don't know exactly how Bill came to this point of view, since I doubt he ever watched TV. Time Warner's announcement of the Full Service Network certainly must have played a role in his thinking. As it happened, the Full Service Network, like Microsoft's efforts in interactive television, was one of the biggest flops in technology history. Since Microsoft lost the Time

Warner bid to Silicon Graphics, Bill decided to align with General Instruments and market to their customers like TCI. At that time, in 1992, all of Microsoft's software ran exclusively on Intel's microprocessors. But these processors were not really designed for the needs of set-top boxes. So Microsoft needed Intel's active support.

Rob Glaser and I established team leaders: Sanjay Parthasarathy from Microsoft and Tim Yiu from Intel. Microsoft had already been in contact with General Instruments, the leading equipment supplier to the cable industry, and wanted them to work with us on the set-top project. Bizarrely enough, GI was led at the time by Don Rumsfeld, the former US congressman and President Gerald Ford's chief of staff and, later, secretary of defense. Rumsfeld was GI's CEO from 1990 to 1993. He eventually became secretary of defense a second time, for President George W. Bush, and oversaw the United States' disastrous war with Iraq.

My first meeting with Rumsfeld was at GI's Chicago headquarters. He sat behind a massive desk, and behind him were large American flags. He was friendly and smiled and joked a lot. He seemed totally clueless about his own business. My point person at GI was its chief technology officer, Matt Miller (no relation). Matt and I became good friends, as we are to this day. Rob Glaser did not seem that engaged in our project, likely because he knew he would soon be leaving Microsoft.

Jerrold Electronics, one of GI's main divisions, was an early supplier to the cable industry. Acquired by GI in 1967 after its 1950 founding, it was sold to Motorola in 2002. The other main supplier, Scientific Atlanta, was formed in 1951 and acquired by Cisco in 2005. The US cable companies split into two camps based on their equipment supplier, finding themselves "locked in" to one vendor or the other because their systems were completely incompatible. For instance, TCI and Comcast were customers of Jerrold. TCI's CEO, John Malone, had actually been head of Jerrold until he joined TCI in 1973. Time Warner and Continental Cablevision were customers of Scientific Atlanta. The cable companies' resulting captivity to one sole supplier would greatly contribute to their thinking when it came to the development of an open cable modem standard.

I pulled together a technical team from within Intel. At that time, Ken Fine was the Intel vice president in charge of embedded microprocessors. Besides its main business selling microprocessors for PCs, Intel also had a robust business in semiconductors for embedded applications, such as telephone switches or military hardware. Ken was also responsible for developing multimedia components, managing Intel's acquisition of DVI from RCA. Ken assigned Tom Franz, a talented engineering manager, to take charge of Intel's activity related to hardware development. I turned to the Intel Architecture Labs for software support, although the majority of the software would come from Microsoft. GI had a semiconductor development group in Arizona, not far from Tom Franz's group in Chandler. Steve Maine, who reported to Matt Miller, managed this group, which worked closely with the Intel team to develop components for the set top. A key element of the needed technology was a set of chips for digital video transmission and decompression, called DigiCipher 2. GI had extensive experience with digital compression, already making products used to transmit digital data by satellite from television networks to cable headends.

At the same time, Broadcom, a company in which Intel had invested through Intel Capital, was developing similar chips. Concerned that GI would not have the chips on schedule, we encouraged them to use the Broadcom chips instead for the set-top box, which they eventually did. These chips were critical for the development of both digital television and cable modems. Intel never considered developing them ourselves.

MONOPOLIST VERSUS MONOPOLIST

When we started to work with the cable industry, I was pretty naïve about the relationships among the various players, especially the role played by John Malone, the CEO of TCI. The joint project among Intel, Microsoft, and GI was started without any involvement of our planned customers, the cable companies. I remember meeting one of the key people at TCI at some industry event. He called us—Intel, Microsoft, and GI—three monopolists trying to take over the cable industry. The cable executives were very aware of what had happened to IBM with respect to the PC. They were not interested in having the same outcome. I could only laugh, since the cable industry itself had a monopoly.

GI planned to supply the set-top box. Intel planned to supply a modified version of the 386 microprocessor with improved multimedia capabilities. The box would provide full interactivity, as well as decompression of digital video. Microsoft would supply the operating system and many of the applications. We estimated the price to the cable companies around $400, but they were looking for something closer to $300. The project was abandoned by GI soon after Matt Miller left the company in August 1994.

At first, I thought that there might be a market for interactive set-top boxes. The more I learned about the cable industry and the more I observed how consumers began to engage interactively with computers in their homes, however, the more the concept of interactive television seemed unlikely. While our interactive set-top box project was not successful, it did keep Microsoft away from AMD. Intel derived great benefit from working on this project: we began to understand the cable companies' infrastructure, technology, and market.

Time Warner's Full Service Network and our Intel/Microsoft/GI project were just two of many projects that consumed billions of dollars in pursuit of a vision of the TV as the interactive device in the home. None of these projects really accounted for the technical and market obstacles. Even now, more than twenty-five years later and no longer facing any technical obstacles, the TV is still not the main interactive device in the home. In 2014, I was an early investor in a start-up, Pluto TV, that took freely available video streams from Internet sites such as YouTube, curated them into channels, and created a passive viewing experience. In 2019, the company was sold to Viacom (for a nice price, may I add). Pluto TV is available on Apple TV and other streaming devices. Today, we interact with our TVs and related devices primarily to select content. It turns out that the "ten-foot experience" is different than the "two-foot experience," especially in the family setting. Nevertheless, even Intel went after the TV experience several more times. In one project, codenamed Hood River, Rob Siegel, who was working for me at the time, devised the concept of designing a PC to fit into the home entertainment center. He got a number of PC manufacturers to join and as a result moved to work for Mike Aymar in the Components Group. While I was not a big fan of the idea, I thought it might be worth an experiment. As it turned out, it was not; the project flopped in the market.

13. RESIDENTIAL BROADBAND

The greatest obstacle slowing the development of the Internet is the limited bandwidth available on the last mile, or the bike path at the end of the information highway as Avram Miller of Intel refers to it.
—Reed Hundt (chairman of the FCC), "The Future of Competition in Communications," remarks to the Washington Research Group, February 2, 1996

Cable companies thought that interactivity would be a tier to their offering, kind of like offering HBO. The cable company would pick the applications that would be available, just as they did with television programming. Fortunately, that idea was defeated by the reality of the Internet. The nature of the Internet was extremely open. Email and messaging really drove the desire to have unfettered access to multiple sites. The cable companies were unable to control access to content.

During the 1990s, large organizations began deploying local area networks, primarily utilizing Ethernet technology, which was 10 megabits per second at the time. Once the Internet began to take off with the 1994 emergence of Netscape, companies and institutions slowly opened up their networks to provide workers with access to the Internet. Portals like Yahoo, Lycos, and even early-stage social networks like GeoCities were now accessible from the office at relatively high speed. I observed employees staying at the office after work to access the Internet, a phenomenon I called "home at work." This informed my thinking on how home computer use would evolve. For those using computers at work, there was no waiting to connect. At the office, computers were always on, always connected, and at high speed. I saw the same future for the home: a new medium for communication, commerce, education, and entertainment. Perhaps it was my intuition at play again, combined with my deep understanding of network technology, that allowed me to formulate this vision. But creating residential broadband would prove a difficult task, one that took almost an entire decade. I would have to work on every aspect of it, including chips, modems, access point, software, applications, and security.

A BRIEF HISTORY OF THE CABLE INDUSTRY

The American cable television industry began in the 1940s. The national television broadcast networks had local broadcasting affiliates that actually transmitted the network broadcast signals in their respective areas, along with local programming. While penetration of televisions climbed rapidly from zero to 80 percent in about two decades, many US homes had difficulty receiving television signals. Unlike radio, the frequency of television broadcasts requires a direct line between the transmitter and the receiver, without terrain, such as mountains and hills, standing in the way. Radio waves, by contrast, are transmitted in a frequency that bounces off the ionosphere, so radio signals can travel up and over terrain. (It is for this reason that you can receive shortwave radio from all over the world, especially at night.)

While television broadcasters would erect towers on hills to reach as many homes as possible, many homes still lacked access to the broadcast TV signals. As a result, several very entrepreneurial individuals began to erect central antennas on hills, amplifying the signals received and distributing them over a fixed coaxial line (that is, a cable) to homes that did not get adequate TV reception. They charged residents an installation fee and a monthly subscription fee. The local TV broadcasters did not mind, as they made their money selling advertising, and cable distribution resulted in more viewers and consequently more advertising revenue.

For many years, cable companies were just small, local businesses. I am sure that these early pioneers did not know that in connecting homes one by one they were laying the bricks of what some would later call the information superhighway. In 1952, there were just fourteen thousand cable subscribers and seventy cable systems. Eventually, some of the cable operators realized they could add original programming. TV sets initially had the capability to receive twelve channels, so there was plenty of space on the coax cable. The first set-top boxes had to be developed to accommodate the new channels. These boxes would convert any channel selected to a fixed channel on the TV (usually channel three). A very important enabling invention was the remote control. The first original program was a weather station in 1956, which was just a camera pointed at a weather station to show temperature, wind direction, wind speed, and humidity. The actual Weather Channel was launched in May 1982. HBO started in 1975, followed by CNN in 1980. In 1970, there were 21 cable-only networks. By 1998, there were 171.

TELEVISION DELIVERY SYSTEMS

By the 1990s, the cable industry faced competition from satellite television companies, like DirecTV and EchoStar, which used a form of digital compression (DBS) and could deliver more channels than the cable companies, which were still using analog technology.

Cable companies received most of their programming by satellite at what was called the "headend," which would have many satellite receivers and served a certain geographic area, perhaps a town or even a neighborhood. From there, they would run coax cable, organizing to reach many homes through a treelike organization of "branch" and "split." Every branch would introduce an amplifier to maintain signal strength, and every split would have some degradation of signal quality. Much of the cost of a cable system was in this infrastructure, termed the "cable plant." The number of channels that could be offered was a function of the quality of the plant, which could be improved by reducing the number of splits or nodes.

In the early 1990s, Time Warner pioneered the deployment of hybrid fiber-coaxial (HFC) network, which was then adopted across the industry. Optical fiber has a very high bandwidth and is immune to noise (electromagnetic interference), so it could intersect the existing cable plant deep within the network, perhaps connecting to as few as five hundred homes, significantly improving quality without the need to replace the cable running into homes. The HFC structure would improve the quality and quantity of television signals, but it was very expensive to deploy, and not all cable companies wanted to make such an investment. A complementary technology, digital compression, could allow multiple television channels to occupy the same bandwidth as one analog channel. Most cable companies used both methods. General Instruments provided the laser systems for transmission.

Both the cable industry and the broader media industry paid great attention to the development of high-resolution television, for which there were many competing strategies. In the computer industry, we wanted to make sure that whatever standards emerged could be easily used on computers. I participated in various government and regulatory discussions on the rollout of HDTV to consumers, including testifying to the House Telecommunications Subcommittee in 1998. The cable industry wanted to make sure the same technology could be used for both digital video compression and HDTV. Simplistically speaking, the idea was that a single analog television channel could be turned into four or more digital channels or one HDTV channel.

TCI, led by Malone, was a collection of low-quality cable franchises all over the United States. Malone decided that rather than invest in upgrading their cable plant, as companies such as Time Warner were doing, TCI would use digital compression. In 1992, Malone announced the "500-channel universe," which he planned to create using digital compression. This was a smart strategy, and it certainly pumped up the stock of TCI and other cable companies. At the heart of its implementation stood two chips, which were also critical elements in the implementation of the set top Intel was developing with GI and Microsoft. As we examined their design at Intel Architecture Labs, we understood that they used network protocols that were packet based and very similar to Ethernet. The chips would provide not only the ability to decompress video but also the connectivity needed to enable interactive service.

BROADBAND VIA CABLE: THE CABLE MODEM

Since 1918, it had been known that multiple analog signals could be combined onto one cable by modulating each separate signal onto its own defined frequency band; the wire could then carry other signals in other frequency bands. On the other side of the wire, the signals could be demodulated and segregated once again. This required much higher bandwidth than was possible with ordinary telephone wire, which led to the utilization of coaxial cable, or coax for short. Coax was invented, like so much else, at Bell Labs, the research arm of AT&T. Invented in 1929, it had a unique capacity for high-frequency bandwidth and could be used over long distances. AT&T used coax to connect cities together starting in the early 1950s, and coax was also used to send television signals between cities. This was the birth of broadband.

> "The set-top box is probably the least important inno-vation," adds Avram Miller, vice president of corporate business development at Intel Corporation in Santa Clara, Calif. "We are going from television where we watch when they want . . . to television where we watch when we want. But you're not going to pay your bills that way."
>
> —"Myths Litter the Information Highway,"
> *Christian Science Monitor*, July 6, 1994

Sometime early in 1993, Matt Miller (GI's CTO) and I met at Intel. I remember borrowing Gordon Moore's personal conference room for the meeting, which also served as the Intel boardroom, because I had something important to share. I told Matt I thought there was no way that the set-top box collaboration among Intel, GI, and Microsoft would ever be successful. We were not getting anywhere close to the cable company's $300 price point. Worse still, the 386 microprocessors, even when specifically modified, were too underpowered for the various applications contemplated. I also shared my view that TVs had insufficient resolution for interactivity, and I questioned the entire concept of interactivity located in the family's entertainment center. Matt agreed; he had thought the same.

Matt and I realized that the cable companies would never pay for a device that could be fully interactive. But I pointed out that consumers were already paying thousands of dollars for personal computers in their homes. Around this time, 25 percent of US homes had computers. We both saw that the future of interactivity in the home would be the PC and not the TV and understood that the chips GI and Broadcom were developing for digital television were packet based, and we had planned on using them for the interactive set top. We asked ourselves: Why could we not develop a cable modem for PCs using these chips? The answer, clearly, was that we could. Matt and I decided to start up a project to do this. We also decided not to tell Microsoft. Later, of course, Bill found out. During one of our joint executive meetings, he screamed at me, "You fucked up everything!" I screamed back, by the way, saying, as I recall, "If you would just shut up, you might learn something!"

Others may disagree, but I never thought that Bill was very strategic or as technical as he likes to present himself. He had bought into the vision of interactive TV because that was what he was being told by the cable industry executives. For Bill, domination was all important. Nathan Myhrvold, Bill's CTO, had developed a comprehensive architecture for implementing the cable industry's vision. They wanted to use a networking technology that the phone companies had developed to integrate voice, television, and data. I believed that all of this could be done incrementally over time using Ethernet technologies. The key here was "incremental." It turns out that I was right, but Bill was angry because doing it this way diminished much of Microsoft's efforts. One such effort led by Craig Mundie was Tiger, which was a way of providing video-on-demand services. Microsoft eventually asked me to publicly support Tiger. Bill and I actually talked to the press together. I was happy to help, but I thought that video on demand was totally missing the point. It would come later with the likes of Netflix and be based on broadband Internet.

To develop cable modems, I had to overcome at least two major obstacles: finding a source of funding and finding engineering talent. Corporate Business Development (later Intel Capital) was basically a venture capital group with a mission to make strategic investments in early-stage companies, not to start internal businesses. I commented to Les that if he provided the funding for the cable modem and related technologies, we would eventually license the technology we developed to other companies in return for stock warrants, thus turning our product-development costs into equity in other companies. This did indeed happen. Finding talent, as it turned out, was not so hard. Intel Architecture Labs up in Oregon was perpetually underfunded, and the failed Gemini Project in 1989 meant they had an abundance of the very kind of network software people we needed. Just as fortunately, Intel had decided to close down a hardware group at its Arizona site, led by Matt Diethelm, that had been part of the Systems Group. The group's employees were looking for jobs elsewhere within Intel. Thanks to support from Les, we kept the group together instead, with a new focus on cable modems and back-end servers.

On July 15, 1993, Matt Miller presented the cable modem plan, called the Edison Project, to CEO Don Rumsfeld and his staff at General Instruments. Matt requested a $2 million initial budget, but whether or not this was approved became unclear. Just one month later, Rumsfeld left his post, replaced by Dan Akerson. Matt took lack of rejection as agreement and began to move the project forward, but he did not get much support from within GI's set-top division.

HYBRID NETWORKS

Not too long after I launched Intel's cable modem project, I was on a flight back to San Francisco from Boston, up in first class, not because Intel paid for first class—cost-conscious Intel most certainly did not, not even to Asia—but because, in those days, if you flew a lot, you'd be upgraded to first somewhat regularly. Sitting next to me was Cathy

Goodrich, now Cathy Lego, a very successful venture capitalist. We had not met before, but we struck up a conversation. She had recently left Oak Investment Partners to go out on her own. She was married at that time to John Goodrich of Wilson Sonsini Goodrich & Rosati, the famous Silicon Valley law firm. Cathy told me about a few of the companies for which she served on the board. Amazingly, one was Hybrid Networks, a start-up working on a cable modem! I had not known about that company, which just shows how random life and business can be.

Hybrid Networks had realized that the biggest limitation to using the cable network for high-speed residential connectivity was the return channel. They had developed a modem that used cable for downstream and a telephone modem for the return upstream, which is why they took the name "Hybrid." They understood that communication between PCs and the network would be fundamentally asymmetrical, such that telephone bandwidth upstream would be sufficient. Hybrid had patented this capability. Soon after that fateful flight, I called Cathy and asked her for an introduction to Hybrid.

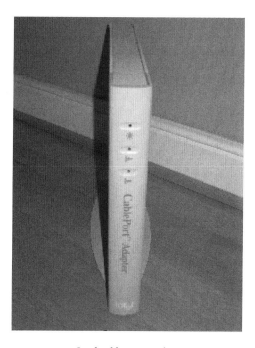

Intel cable port modem

Hybrid, started by Howard Strachman and Ed Moura in 1989, was quite visionary in realizing the potential held by the cable system for high-speed residential communications. When I met them, they had a prototype, a stand-alone box that connected on one side to both the cable network via coax and to the telephone system via a standard phone cable, while connecting to the computer through an Ethernet port. Though Ethernet controllers were not common on PCs of the time, requiring the installation of an add-on board inside the box, neither of the only two typical external ports on PCs, the serial and parallel (printer) ports, could receive high-speed data. They had developed software to combine the downstream and upstream within their box and send it to a PC via Ethernet.

For testing, Hybrid had obtained use of an abandoned television channel (UHF), with a transmitter located on a mountain top in Silicon Valley. A high-speed T-1 telephone connection (about 1.5 megabits) ran from their office to the transmitter, which would then broadcast the data. The return path, as I mentioned, was by ordinary telephone line to Hybrid's office.

I met with Hybrid Networks early in 1993, when the Internet was just starting to become accessible. Mosaic had just been released; Netscape was a year away. Hybrid demonstrated the possibility of high-speed communication using the Internet. For instance, they connected to a government website with very high-resolution weather maps. They downloaded almost instantly, faster even than loading from a CD-ROM.

Hybrid to set up a demo at Intel so I could show both Andy and Les. We set up in a conference room with line of sight to Hybrid's UHF television transmitter. Andy and Les were very impressed. In November 1993, we lent Hybrid $2 million in convertible debt and entered into a licensing agreement. In 1995, we converted the debt into equity, ending up as the largest investors in the company (more than 12 percent) after it went public in 1997.

I asked Hybrid to set up a connection at my home on Emerson Street in Palo Alto. By that time, Intel had relationships with both AOL and Prodigy, so I arranged to connect to these services over the Hybrid high-speed service. I am sure that was the first high-speed wireless connection from anyone's home to those online services. To accomplish this, Hybrid had to put a UHF antenna on the roof of my home. I remember thinking that my neighbors might think I was some kind of spy.

While Hybrid's approach (cable down and telco up) had some merit, I did not see it as the real solution. I wanted a modem that required only a single connection for both directions: to the cable television network. One thing that had become very clear to me was that creating a new medium would require a network that was very fast, always on, and always connected. I was very excited about the possibilities for cable modems, convinced that this would be the beginning of something really big. I was right; it was!

A symmetrical network was not required; the desired speeds of data transfer from the network to the PC far exceeded the speeds needed to transfer information from the PC to the network. Most of the time, I realized, the needed upstream speed would be bounded by how fast a user could type or click. That was very good news, since the upstream return channel in most cable plants was a disaster and was allocated only a small amount of bandwidth. The cable companies put in amplifiers and filters at each of the nodes they served, sometimes even on the homes themselves. While the cable specifications provided for a small upstream return path, with the envisioned use of sending program selections back to the cable provider, that path was never really used. In fact, many cable companies had disabled the upstream altogether. Even when enabled, the upstream channels were incredibly noisy, one of the greatest impediments we faced in deploying broadband cable.

I brought General Instruments into the deal with Hybrid. Matt and I both saw that using Hybrid's technology would let us move ahead rapidly. We did not think Hybrid would provide the cable modems themselves over the long-term. Instead, their business would migrate, we thought, into providing the back-end servers and licensing the technology, since both Intel and General Instruments could produce consumer products at higher volume and lower cost. I introduced the company to Accel and Sequoia, two leading venture capital firms, which both invested. AT&T also joined.

While Hybrid's very early support of Intel's cable-modem activities was crucial, Hybrid ultimately did not continue as a supplier of cable modems or headend equipment. Standardization efforts would commoditize such products, and a small company like Hybrid would not be able to compete on price. Instead, Hybrid focused on what it called wireless cable modems, which returned to the idea of using TV broadcast channels and telephone return. That business did well for a while, and they even had a deal with telecom Sprint. They eventually ran out of money, however, and their major debt holder acquired the assets, including the intellectual property. Hybrid also faced a number of shareholder lawsuits regarding the overstatement of revenues at the time of their IPO, and the SEC fined three Hybrid employees, including CEO Carl Ledbetter, who had been brought in from a senior position at AT&T to replace Howard Strachman in 1996.

WHY CABLE COULD OFFER BROADBAND INTERNET

Each television channel had 6 megahertz of bandwidth, which made it possible to have hundreds of channels on one cable. Already in development were plans to use one analog television channel for four digital channels with identical resolution and quality. We decided to use the same technology for the cable modem, which could give us as much as 30 megabits per second of downstream data. The upstream bandwidth would be much more limited, but it would be sufficient for many customers.

Cable utilized a shared connection, with multiple homes on the same cable, similar to Ethernet when it first was deployed. The bandwidth needed per home, however, could be easily accommodated. As more and more homes were added, the number of homes on a single shared node could be reduced by deploying more fiber optics into a hybrid fiber-coaxial network. The nature of Internet use initially involved a human being on one side and a fast computer network on the other, meaning that activity was intermittent, even with multiple people on the network simultaneously. Later, such technologies as streaming video and online storage would change that paradigm.

As I learned more and more about the cable infrastructure, I became increasingly excited. I thought the achievable bandwidth would be pretty much unlimited. We started with 10 megabits per second, but a full 30 megabits was our target, equivalent to one TV channel. I also thought that the cable companies could allocate more than one channel to data service, as there were so many channels on the coax line, increasing speeds. It also seemed possible that ways could be found to increase the number of bits per second per channel. All of this happened eventually. As I write this in 2021, almost thirty years later, downstream cable modem speeds have reached 10 gigabits per second, and upstream speeds are 6 gigabits. This is more than a thousand times faster than the first cable modems.

It is important to recall that the cable companies had no idea they were sitting on oil fields. They still thought they were in the television business, and they hoped that they might also get into the telephone business eventually.

CASTRO VALLEY VIACOM TRIAL

Time Warner, together with Silicon Graphics, was in the process of developing their Full Service Network, which would be trialed in Orlando, Florida, from 1994 to 1997. While exploring implementing a small-scale version of the Full Service Network, Doug Semon, director of Viacom's cable operation, had put out a request for proposal (RFP) for the various pieces of technology needed. Hybrid Networks had sent a proposal using their cable modem that at first did not make sense to him, because it did not concern interactive television at all.

Doug was a bit geeky, having assembled his own computer out of parts. He must have understood the potential for the cable industry if it offered broadband Internet. He met with Hybrid's founders, Howard Strachman and Ed Moura, who demonstrated a cable-based connection to one of the few web servers in the world. Very excited at the potential to provide high-speed connectivity to personal computers, Doug sought support from his management to do a trial.

When Hybrid told Intel and GI about their discussions with Doug, we saw Viacom as an excellent opportunity to do a trial with actual consumers. Getting Viacom's management to agree took a while, even involving discussions with their CEO, Frank Biondi.

As I mentioned above, I was convinced we needed a modem that worked bidirectionally on cable, without requiring an upstream telephone line. PCs, I thought, would have to have an always-on, always-connected, high-speed communications capability. A very talented Intel engineer named Lew Adams modified the Hybrid modem to use the upstream capability of Viacom's cable plant.

The trial took place in Castro Valley, California, a community east of San Francisco Bay. The first modem was installed in October 1994, but no one was charged until December 1994. We started with 50 homes, which grew to 256 by June 1995. The downstream rate was 10 megabits per second. While the upstream rate was just 128 kilobits per second, state-of-the-art telephone modems of the time were limited to 28.8 kilobits per second. Viacom and Intel conducted focus groups. Users were extremely positive, which was very important in gaining support not only within both Intel and GI but also, critically, the other cable companies. One man with family in East Asia, when asked how the modem had made a difference in his daily life, literally had to choke back tears at the question. He could now communicate with his loved ones in an unprecedented way. "This thing has changed my life!" he said, losing his composure completely five words into his answer. Many users even refused to return their cable modems at the end of the trial.

During the same period, our Intel/GI/Hybrid partnership did another very successful trial with Comcast in Lower Merion Township, Pennsylvania. Many years later, Doug and his counterpart at Comcast, Steve Craddock, argued with each other by email about who was first. To my mind, they both were.

COMCAST

The story about how we ended up doing a trial with Comcast is a bit more convoluted.

In early 1993, Andy Grove called to tell me that a journalist friend of his, Norm Pearlstine, wanted to learn more about the Internet. Andy thought I could be helpful to him, so he had recommended Norm call me, as he did shortly after that. We met at his hotel in San Francisco a few days later. In his suite, I spoke for a while about computing and offered some of my thoughts about how computing in the home would evolve. Norman was very good friends with Barry Diller, he told me, who had recently left Fox, a television network he had created for Rupert Murdoch. Barry had invested $25 million of his own money into a shopping network, QVC, headquartered close to Philadelphia and part owned by both Comcast and TCI. Norm thought Barry would be interested in hearing my views about home computing, especially the potential development of residential broadband.

Barry Diller's office set up the meeting. My boss, Les Vadasz, joined me in traveling to Philadelphia. QVC was an amazing operation, though I frankly sensed that the typical QVC customer would not be an early adopter of residential broadband technologies. Out of that first meeting, Barry and I developed a friendship. I visited with him from time to time at his home in the Beverly Hills. (I also often met him and his wife, Diane von Fürstenberg, at the Allen & Company media conference in Sun Valley, which I attended from 1993 to 1999.)

In speaking with Barry at QVC, he said, "You know who you should talk to? You need to speak with Brian Roberts at Comcast!" He asked if we had time to meet with Brian. When we told him that we did, Barry got on the phone. An hour later, Les and I were at Comcast, where we met with Brian Roberts and his father, Ralph. Though Ralph was still CEO, when it came to technology, he deferred to Brian, clearly the heir apparent. Brian was very interested in the potential of using his cable network to connect PCs at high speed.

I had already begun discussions with Comcast, as I, of course, told Brian. I had already met Mark Coblitz, vice president of strategy, at a Kagan conference, and we had discussed cable-based broadband. Mark is a very nice, very bright man who played a big role in influencing Comcast to embrace broadband connections to personal computers.

Intel and Comcast reached an agreement to conduct a five-hundred-home trial in Marion, Pennsylvania. I was really taken with both Brian and his father, Ralph, who actually shared the same office, with two desks side by side. Brian, more than anyone else I ever met in the cable industry, truly saw the full potential of providing access to personal computers. Over the years, I became friendly with the Roberts family. Brian eventually asked me if I would be interested in joining Comcast to start a corporate venture group. I joked that I would only consider it if Ralph adopted me, and I turned down the offer for a number of reasons. Once Comcast developed into the powerhouse it is today, I had some regret about that decision. After I left Intel, Brian and I spoke some about my joining the Comcast board. Though I would have liked that, it did not happen.

> Years ago, Miller identified cable companies as key players in connecting home PCs to the Internet—and therefore connecting people to each other. At his urging, Intel forged early development of cable modems, which allow PCs to send and receive information much faster than standard telephone modems. The first cable modems are expected to hit the market later this year. "He deserves much of the credit for their development," says Brian Roberts, president of cable company Comcast.
>
> —Julie Schmit, "For Intel, He's a One-Man Think Tank. He's Also Plugged into Hollywood." *USA Today*, March 14, 1996

PITCHING TO THE CABLE COMPANIES

As the primary leader of the Intel cable-modem consortium with Hybrid and GI, my role was to get both the cable and computer industries to buy into the concept of a broadband capability for residential computers, not such an easy task. The cable industry thought of itself as being in the television business, after all. And so I became an evangelist. I met with the CEO of every major cable company around the world. The United States had an unusually strong cable infrastructure, but there were many other countries with well-developed cable industries, including the United Kingdom, Holland, France, Japan, and Australia.

Brian Roberts of Comcast and John Malone of TCI were extremely important in our deployment. Their companies were also the two most important customers of General Instruments. While I met the CEOs of Time Warner Cable, Continental, and MediaOne, they were strongly in the Scientific Atlanta camp of set-top customers, so our relationship with General Instruments got in the way, as did Time Warner's doomed love affair with the Full Service Network.

In the summer of 1993, I attended the most important annual conference for cable executives held that year in Bermuda and attended by executives from the top two hundred cable companies. It was called A Vision for the Future and was sponsored by Daniels & Associates, the leading investment bank for the cable industry and for General Instruments.

— Don't use tomorrow's technology to
implement yesterday's vision.
—Avram Miller, Daniels
Conference, Bermuda, 1993

There, I had the opportunity to introduce Steve Case, AOL's CEO, to Scott Kurnit, the EVP of marketing at Prodigy. I was friendly with both of them. We all had a sense of humor, so we had a good time.

In my talk at the conference, I made three points. Keep in mind, this was before companies such as Yahoo, Amazon, or Netscape were founded.

1. Consumers are already learning interactivity at work. Over half of white-collar workers use a computer every day, and we expect near-complete penetration by the end of 1999. There are already 22 million computers in US homes, and we expect six million or more to be purchased in the next year. The average consumer spends about $2,000 on their computer. Consumers spent more

on sound cards for their home computers than for their set-top boxes.

2. Any development of interactive TV would have to stand on the back of the computer industry, because the investments required otherwise would be prohibitive. The computer industry is already digital, networked, and interactive.

3. While many of you think that interactivity is all about entertainment, I think it is all about software. We don't know what the applications of interactivity will be. We need to reduce barriers to entry and enable early-stage companies to create the application software.

On February 15, 1994, we organized a meeting with key cable industry players, GI, and Hybrid Networks. Bruce Ravenel attended for TCI, and Mario Vecchi and Carl Rossetti attended for Time Warner. Though still busy with the Full Service Network, Time Warner was starting to think about how PCs might fit in. Ravenel admitted that he had changed his mind about PCs connecting to the cable network; he now thought it important. We discussed, at this meeting, the need to have an open system based on standards, and we considered a joint venture to develop the needed specifications. (I think after this meeting, Ravenel may have convinced Malone to repurpose the MCNS partnership, as described below, for cable modems, but I am not sure. Though there were follow-up meetings, I only have notes from the first one.)

After the meeting, Mario Vecchi wrote a memo to his boss, the CTO of Time Warner, Jim Chiddix. He understood very clearly what we were trying to do. "From the PC user's perspective," he wrote, "it will pretty much look like an Ethernet connection to remote information services using TCP/IP networking." Mario was able to convince Time Warner executives to pursue, in his words, the "quite attractive" market Intel had described, which led to Mario's work co-founding Time Warner's RoadRunner broadband service.

EARLY CABLE MODEM TRIALS

Both Viacom and Comcast announced in advance of the Western Cable Show in December 1993 that they would start their respective cable modem trials working with Intel, GI, and Hybrid. At the Western Cable Show in Anaheim, California, Intel demonstrated our modified Hybrid cable modem in the GI booth, which used the cable network for upstream. LANcity and Zenith had also developed cable modems by then; Zenith showed theirs in their booth, which focused on set-top boxes. I don't believe LANcity was present. A consortium of Cox Communications, Prodigy, and Zenith also announced residential trials at the show, but reliability problems with the Zenith modem kept that group from going very far. Continental Cablevision, LANcity, and Performance Systems International (later PSINet, one of the first Internet service providers) fielded a pilot in Massachusetts. Each group offered incompatible solutions to the cable industry.

I remember so many people coming up to the GI booth at the 1993 Western Cable Show and wondering why we had a computer there. Just one year later, at the 1994 Western Cable Show, there were many cable modem vendors, and Intel itself had its own booth. Within a few more years, there would be more than twenty manufacturers of cable modems.

The Western Cable Show was fascinating. There were actually two exhibit floors, one for equipment manufacturers, like General Instruments and Scientific Atlanta, and one for television networks, like CNN and HBO. Before the major upgrade of the cable system over the second half of the 1990s, new cable television networks had to work hard to get "carriage" over the cable systems. Not that many channels were available on the cable plant. With literally hundreds of television networks seeking a place on cable, by the beginning of 1990, fewer than eighty could be transmitted. In less than a decade, that number would double. In the television network exhibits of the WCS, you could find religious channels next to adult channels and legal channels next to crime channels.

While the booths of the cable equipment companies were filled with businessmen in suits, the networks staffed their booths with attractive young women.

In terms of content, when we started our 1993 trials of residential broadband, we had already announced agreements with Prodigy, AOL, and several other online services, but they had not designed their content for high speed. So, we had created what we called a "CD carousel," because the only rich, multimedia content available for PCs at that time was on CD-ROMs. The CD carousel was metaphorical, allowing the user to download a CD in about the same time as they could have from a local CD player to access content such as encyclopedias, maps, and games. Intel well understood that rolling out residential broadband would require a great deal more than just developing cable modems and upgrading the cable infrastructure. We also had to make sure there would be compelling content that needed high-speed access. Intel played this very important facilitation role well, especially by making investments in early-stage companies through Intel Capital. For instance, we invested in Launch Media, which produced music-related multimedia content on CD-ROMs. As a condition of our investment, we required that Launch focus on the Internet. Launch was later acquired by Yahoo and Yahoo Music.

EARLY DAYS OF CABLE BROADBAND

There had been several early attempts to use the cable infrastructure to provide interactive service, the best known of which is probably QUBE (1982), which I described above. Sytek, a networking company, in collaboration with General Instruments, developed a cable-based broadband system for both televisions and personal computers that was compatible with existing cable networks. For a number of reasons, including expense, it was never commercialized.

Many companies, such as Ungermann-Bass, developed networking products that used broadband on coaxial cable, but these were not compatible with existing cable networks. They were really competitors to Ethernet: alternative ways of implementing local area networks.

Zenith developed a series of products that at first targeted manufacturing facilities. Ed Zylka was the key developer, and he later moved to work for GI. Initially, Zenith's products concentrated signals from terminals, but they then moved to support Ethernet. Out of this activity, Zenith eventually created a proper cable modem, branded HomeWorks, which allowed a personal computer to communicate with network services using a confidential, residential cable system. They did a residential trial with Prodigy and Cox Communications, as I mentioned above.

David Fellows, the CTO of Continental Cablevision, was one of the savviest technologists in the cable industry at that time in terms of computer networks, and he was perhaps the first to recognize the Internet's potential. When he saw a demo of Mosaic, the forerunner to the Netscape browser, Fellows thought that its point-and-click aspects had potential for use in interactive television. As it happened, he played a key role in creating cable-based residential broadband. He reached out to Delphi in Cambridge, Massachusetts, one of the early Internet service providers, to discuss the possibilities of using the Internet for interactive TV. The engineers at Delphi convinced Fellows to focus on home PCs, rather than interactive TVs.

Continental, based primarily on the East Coast, was working with a small company in Andover, Massachusetts, named LANcity. Its founder, Rouzbeh Yassini, is one of those who likes to refer to himself as the father of the cable modem. Though Rouzbeh made many contributions to the development of cable modems, particularly to their standardization, he tends to overstate the importance of his role, in my opinion.

LANcity's business at that time was to connect an enterprise's facilities together using cable, basically creating a bridge between standard Ethernet on one side and a coaxial cable on the other. Paired with another bridge, this would allow two LANs to be connected at relatively large distances. For a time, for instance, Digital Equipment Corporation used LANcity equipment to connect networks at two of their sites together. This concept was termed a "metropolitan access network" (MAN). Their first-generation product cost about $20,000 per unit; a second-generation unit cost about half that. LANcity would develop a product at a consumer price point by the time they were acquired by Bay Networks in 1996, though it remained at a symmetrical 10 megabits per second in bandwidth. This meant it was considerably slower compared to the asymmetrical market leaders in the downstream bandwidth that most mattered to consumers. Because their technology was still perfect for testing consumer reactions to high-speed connectivity, however, their contributions were very important. I still don't understand, frankly, how they achieved 10 megabits per second in upstream bandwidth at that time. Very possibly, their cable partners simply overbuilt their plants to accommodate the wider upstream channel, adding extra coax for this purpose. From Intel's perspective, we understood that the kinds of volumes we needed to expand the size of the PC market would only be possible with a modem that could access a typical cable plant.

At Continental, Fellows found a company, Performance Systems International (PSI), that could connect to the nascent Internet. Continental and PSI created a joint venture and bought a bunch of "cable modems" from LANcity, which were reduced-cost versions of their symmetrical MAN product. Their announcement got a lot of press, including an August 30, 1993 article in the *Wall Street Journal*. I believe this was the first major public discussion of cable modems.

A group of cable companies, including Continental Cablevision and MediaOne, partnered in 1995 to create Highway 1 using LANcity modems. To my knowledge, they ordered five hundred thousand modems, which certainly made LANcity the leading supplier of cable modems at that time. MediaOne's broadband service, MediaOne Express, was co-branded with Time Warner as MediaOne RoadRunner. Since I don't believe these LANcity modems could have been compatible with the cables carrying video, I think they must have overlaid a second cable system to each home served.

In 1996, MediaOne acquired Continental Cablevision, and Fellows became CTO of MediaOne. In 1999, AT&T, which had already acquired TCI, bought MediaOne, making AT&T the largest cable company in the world, and Fellows became CTO of AT&T Broadband. Later, AT&T would sell its cable business to Comcast, and Fellows then became CTO of Comcast. Across Dave's remarkable career, he made many contributions to the development and growth of residential broadband.

Cable modem by General Instruments

Meanwhile, Intel and GI used Hybrid Networks' cable modem technology that could use existing cable plants. As I described above, Hybrid had at first used telephone connections for upstream, since the upstream portion of most cable systems was either disabled or too noisy to be used. Intel modified Hybrid's product to provide two-way communication over cable. We used the resulting product in trials, as I described above, but Intel would not bring it to market.

Intel and General Instruments used some of Hybrid's technology to build a cable modem that could be installed directly in PCs. General Instruments then developed a stand-alone modem that connected to a PC by Ethernet, the forerunner of the cable modems to come. Ed Zylka named the product the SURFboard, because many of his engineering colleagues in San Diego were part of the surfer culture there. General Instruments would eventually become the leading supplier. Motorola entered the market in 1996 and became the second-largest supplier. In 2002, Motorola acquired General Instruments, replacing their line of cable modems with that of GI. Finally, in one of the most bizarre series of acquisitions I can recall, Google acquired Motorola in 2012 and then sold the cable modem business to Arris, which, incredibly, was later acquired by CommScope, itself originally owned by

General Instruments. As for LANcity, Bay Networks acquired them in September 1996, and Nortel acquired Bay in 1998. Nortel went bankrupt in 2009, and Arris also acquired Nortel's cable modem business. Arris cable modems still carry the SURFboard name, tracing their ancestry right back to the modem Ed Zylka developed at GI after he left Zenith.

By 1995, the Intel relationship with General Instruments was not going smoothly. Their set-top business was located in Pennsylvania with the Jerrold Division. Hal Krisberg, the general manager of GI's set-top business, was not that keen on working with Intel. Our main champion, Matt Miller, was having his own difficulties with then CEO, Dan Akerson. We decided to go our own separate ways with the development of cable modems.

In 1987, GI had acquired the part of M/A-COM that had an encryption technology called VideoCipher. This, it turned out, was an extremely important and strategic acquisition, because the follow-on technology, DigiCipher 2, eventually led to the development of HDTV and digital compression of conventional television, further helping to set the standards for digital transmission of data by cable. General Instruments had combined both the Jerrold and VideoCipher Divisions into the General Instruments Communications Division, under Hal Krisberg's leadership.

VideoCipher had originally been developed as part of a company in San Diego called Linkabit by the same people who went on to found Qualcomm, a leading digital cellular technology company. The two parts of the division were not only separated by a continent but also had totally different cultures. Moreover, Hal did not really understand the cable modem opportunity. He focused primarily on the development of the next generation of cable television boxes, which would support digitally compressed channels. Hal also wanted these boxes to provide telephone services. Meanwhile, the management of the San Diego part of the division was also not crazy about working with Intel; because their core competency was in semiconductors, they worried that Intel would get into their business once we understood how the technology worked. According to Mike Ozburn, who worked there, GI was also very focused on short-term sales, and the cable modem business would clearly take some time to develop into a significant business.

Motorola entered the market toward the end of 1996, first targeting Zenith and then going after the entire market. By the end of 1996, they had only shipped fifty thousand units. Like everyone else, they had proprietary technology that was incompatible with any other modem in the market. Leading Motorola's efforts was Jim Phillips. Motorola had deep knowledge and experience with radio-frequency technology because of their leadership position in cellular phones, and they decided that the cable industry could be a strategic market for them. They called their first cable modem product CyberSURFR.

I first met Jim Phillips at the 1997 Allen & Company Sun Valley Conference. He claimed such high sales numbers that I thought he was exaggerating. He was not. Over time, the smaller players gave up competing.

Intel decided to exit the cable modem business in April 1997 once it became clear that it would be a commodity business with little opportunity for profit. Frankly, I was happy about this; I wanted as many manufacturers as possible of cable modems in order to drive down the price, thus serving Intel's primary goal of making the PC the home's primary interactive device. The fact that Intel no longer offered a cable modem made it easier for us to work with all the other vendors. Matt Diethelm, who led Intel's cable modem group, retired. His team of seven transferred to the Intel Architecture Labs, reporting to Ali Sarabi, my main collaborator there. Almost twenty years later, in 2010, Intel would acquire the Texas Instruments broadband division and become the leading supplier of components for cable modems, long after GI was gone, only to exit from this business again ten years after that.

14. CREATING THE STANDARD

The cable companies established CableLabs in 1988, as an independent nonprofit technical organization to serve their industry. Richard (Dick) Green was its CEO from 1988 to 2008, when he retired. I was fortunate to have the opportunity to work with Dick, who played a very important role in the cable industry's development, particularly in the area of standardization. One benefit of CableLabs was that its board comprised the CEOs of all the major cable companies, allowing them to gather together often. I am sure they discussed matters besides standards and technology. John Malone was the board's dominant personality, both because he was CEO of the giant TCI and because he was the most technical of all the CEOs, having started his career at Bell Labs.

Dick Green organized one trip every year for cable industry executives to meet with executives from other industries. For instance, one year, they all flew to Japan to meet with executives of the consumer electronics industry. In 1996, they paid a visit to a few Silicon Valley companies, plus Microsoft. They were interested in learning from the tech executives, and they wanted to convince them that cable was indeed the best way to offer high-speed residential Internet. They brought along a presentation on the industry's plans for the upgrades needed to establish cable-based broadband.

I organized the visit at Intel. Andy Grove attended. He was not the most popular person with the cable industry at that time. In July, he had taken a public position at the Allen & Company conference that the cable industry was "too old and feeble" to bring broadband into homes. Cable industry stocks were badly hit as a result. It never really became clear to me whether Andy really thought that or if he thought that by saying those things, he would, through "management by harassment," force the cable industry to act more aggressively. His behavior certainly did not help my ongoing discussions with cable industry executives. I think Andy was impressed after meeting with them, because after that, he stopped denouncing the cable industry to the press.

After our meeting, the cable executives flew to Redmond and had dinner with Bill Gates. Gates told the cable CEOs that he was impressed with the potential of residential broadband on cable. Brian Roberts of Comcast, just thirty-six years old, half joked that if Bill was so impressed, he should buy 10 percent of all the cable companies. Bill asked, seriously, how much that would cost. The cable companies were worth about $50 billion at that time, close to the value of Microsoft, so the answer was $5 billion. Bill had $10 billion in cash, he said, so he could do it. It still seemed like a joke. The next day, Greg Maffei from Microsoft contacted Brian Roberts. In June 1997, Microsoft purchased 11 percent of Comcast for $1 billion and then promoted Maffei from vice president to CFO. The investment had a greater impact than just the money itself. A shot in the arm, it did much to rectify the financial damage resulting from Andy Grove's comments. The day after Microsoft's investment, the index of cable company stocks climbed 13 percent.

Microsoft was very interested in providing products to the cable industry. Their numerous programs included WebTV, Tiger (a video-on-demand server), and an operating system for set-top boxes, but none of these products were very successful. Microsoft's benefit, just like Intel's, would come from growth in the market for home computers.

IF IT SEEMS TOO GOOD TO BE
TRUE, IT PROBABLY IS

A new microprocessor company, MicroUnity, claimed a breakthrough in microprocessor architecture that could totally revolutionize set-top boxes and cable modems. Many cable companies were interested and wanted to invest. To put it mildly, this was fantasy, but many cable companies believed the claims. Around 1996, led as usual by Malone, the cable CEOs formed a partnership, MCNS (Multimedia Cable Network System Partners, Ltd.), to manage the cable companies' investment in MicroUnity. Microsoft and Hewlett-Packard invested as well. Some called it The Cray in the Tray, meaning a supercomputer in a set-top box. It took about two years, but eventually they all came to their senses; MicroUnity would never be able to meet their requirements. By this time, though, Malone had realized cable modems' potential, and he redefined the role of MCNS to focus on cable modem standards. MicroUnity, meanwhile, turned more or less into a patent troll. They sued indiscriminately around the industry, including Intel. I actually had to give a deposition. Intel eventually settled for $2.2 billion dollars, I believe just so we'd not have to deal with them any longer. The promise of MicroUnity was that their technology would dramatically increase the capabilities of set-top boxes while not increasing their costs significantly. With the failure of the Time Warner Full Service Network in 1997, it had become apparent that offering interactive television was not a viable business, while at the same time, the growth of the home computer market created an opportunity for the cable industry to become broadband Internet service providers.

The cable industry had found itself locked into sole-source vendor relationships with companies like General Instruments and Scientific Atlanta that used proprietary, incompatible technologies. For instance, if Time Warner had a franchise in Los Angeles and they wanted to buy additional set-top boxes, they would have to purchase them from their existing vendor, Scientific Atlanta, or replace all set-top boxes across their entire system. Once the cable companies, and especially Malone, began to understand the potential of residential broadband, they wanted to make sure that they would not be locked in. Malone convinced the others to redirect the work of MCNS to develop standards for cable modems. Interestingly, Malone had formerly been in charge of the Jerrold Division of General Instruments and had certainly worked hard at GI to lock his customers into GI's technology. He did not want to see this repeated now that he was on the other side. On the other hand, cable modem manufacturers like Motorola and General Instruments were not happy with the idea of a cable industry standard (and one controlled by that industry). They would have preferred to have preparatory solutions to prevent low-cost manufacturers from competing.

Intel had a different perspective. While we had toyed with the idea of having a cable modem business, such a business would have been small and would not have been that profitable. We had already been working on standardization along with HP, AT&T (and later, its spin-off, Lucent), and Hybrid. Our group was called the Broadband Link Team. Intel engineer Lew Adams was co-chair of this group, and he personally made many contributions to what eventually would become the cable modem standard. Our main interest was seeing more PCs in homes. We could easily have given up our desire to start a cable modem business and instead give our technology to MCNS.

MCNS hired Arthur D. Little to work with the leading cable modem vendors to pull together a standard. Simultaneously, Motorola, Bay Networks (which had acquired LANcity), and many other potential vendors were also working on their own proprietary standards. Eventually, the Broadband Link Team decided to transfer all of their work to MCNS, and the other various companies abandoned their proprietary standards. The standard that came out of MCNS was called DOCSIS, which stood for data over cable service interface specification.

It remains the cable modem standard. More than a billion DOCSIS-compatible modems have been produced to date. The standard has been updated several times, adding both features and increased bandwidth. As I write this, we are now at DOCSIS 4.0. DOCSIS has been an amazing accomplishment.

Key cable-system members of MCNS included Rogers Communications, Continental Cablevision, TCI, Time Warner Cable, Comcast, and Cox Communications. In terms of vendors, General Instruments worked with MCNS, as did Com21, Hewlett-Packard, and Motorola. Far and away, the company with the most impact, however, was Broadcom, which made key components needed by both cable modems and digital set-top boxes. A number of cable participants have credited Tom Quigley of Broadcom for practically writing the first DOCSIS specification for themselves. The CEO of MCNS was Stephen Dukes, also vice president of TCI Technology Ventures, led by Bruce Ravenel, who will come up again, later in this story, as one of the key people behind @Home.

Once the MCNS consortium approved the initial specification, it was transferred to CableLabs, which then formalized the specification and developed the certification process. Robert Cruickshank III led this work, and explained the development process in detail in a book much later published by the Cable Center, *The Souls of DOCSIS*. CableLabs put out a request for proposal in early 1995 outlining the cable industry's requirements. In all, fifty companies responded by the June deadline, including many already mentioned. Some were vendors to the telephone industry, offering proposals that would time slice the bandwidth so that all homes got a small piece of bandwidth, even if they did not need it. Fortunately, the vendors that were already working with the cable industry and doing trials based their solutions on the Internet model, according to which bandwidth is largely allocated on demand. As a critical requirement, the CableLabs team also had to figure out how to test and certify cable modems as interoperable.

It was an amazing effort by many companies and by CableLabs, which released DOCSIS 1.0 in 1997. After LANcity was sold to Bay Networks in the middle of 1996, CableLabs hired the founder Rouzbeh Yassini to oversee the certification process. Yassini, like many others, played a major role in the development of residential broadband. He describes his story and vision, along with others' contributions, in a 2003 book published by Cisco Press, *Planet Broadband*, which he wrote with several cable industry journalists and analysts.

The first DOCSIS-compatible modems came to market at the end of 1997, with pricing below $200 per unit. The first DOCSIS modems were 40 megabits down and 10 megabits up. Regular updates to the specification and the underlining technology have brought possible speeds, as of DOCSIS 4.0, released in 2020, to 10 gigabits down and 6 gigabits up. Though still a shared medium, this is now so much bandwidth that it allows multiple streams of HD video inside one home, providing the foundation for streaming video companies like Netflix and Hulu.

DOCSIS 1.0 only supported external modems, and Intel decided it was strategically important to get the specification to support internal modems, as well, busy as it was trying to push as much functionality directly into the PC as possible. An internal cable modem would have greatly reduced cost, eliminating the power supply, external casing, and much of the electronics. CableLabs required Intel to actually build such a product in order to have it verified and added to the specification. We did so, but an internal cable modem attracted no market interest, because installation would require opening the computer case. Consumers did not want cable technicians, who they really did not trust, cracking open their computers. Home PCs did not have Ethernet, but it ended up becoming a lot easier to get Ethernet installed than to get internal cable modems installed. Soon after, USB would start to be included with all PCs, and external cable modems would be sold with both USB and Ethernet connectivity. Eventually, cable modems would add router functions and Wi-Fi, as homes had more and more computers and other devices.

I was, of course, delighted with the DOCSIS standardization effort, as having many companies compete to provide cable modems was in Intel's best interest. We wanted the retail market to have low-cost, readily available modems. Early cable modems, including trials by the Intel/GI/Hybrid consortium, Zenith, LANcity, and Motorola, helped get the cable industry crucial experience that allowed DOCSIS to be completed successfully. Since the vendors actually understood more at first about cable broadband than CableLabs, the DOCSIS team wisely chose to work closely with a number of key vendors.

A NEW BUSINESS MODEL FOR CABLE

I remember an early discussion I had with John Malone in his office outside of Denver, Colorado, sometime in early 1994 about the potential of broadband. I suggested that cable modems be sold through retail channels. John said emphatically that he would never let his customers buy their own cable modems. As with set-top boxes, he insisted, TCI would own the modems and rent them out to their customers. He also wanted to completely control all the hardware. I explained that if he allowed cable modems to be sold through retail channels instead, along with a subscription to the cable online service, he would end up with a new sales force. That, he seemed to understand. John was, above all else, very bright and very greedy. The problem with retail sales initially was there was no way to connect cable modems to personal computers without installing an Ethernet card. This would change as Ethernet cards came bundled in and especially after the computer industry, led by Intel, developed USB and worked to get USB part of the cable modem standard.

Every cable franchise was a closed system. You could watch only whatever television programming your cable company provided. Recall that at the time of the early cable modem trials, online service companies, too, such as AOL and Prodigy, followed a "walled garden" model. At first, cable companies imagined that all online services would be offered by them just like they offered HBO. Comcast actually approached AOL with this model but was turned down. For the longest time, online service companies such as AOL also had a "walled garden" approach. It may be hard for those who don't remember that time to believe, but all the content that we got was curated. You read what the editors of a newspaper or magazine chose to print. You watched the shows on TV that NBC, CBS, or ABC offered you. Throughout the '90s, Internet companies like Yahoo, Excite, and Lycos envisioned that their portals would be like a television network. But that turns out not to be what consumers wanted. So like a hole in a dam, the Internet let out all the water.

For instance, Time Warner, which was primarily a media house with cable network, wanted to create their own portal to the Internet, as portals were thought at the time to be consumers' primary point of contact with the web. In January 1994, Walter Isaacson, then editor of new media for *Time* magazine, wrote a memo to Jerry Levin, Time Warner's CEO, suggesting that Time Warner integrate all the content from the various Time Warner divisions into one comprehensive site. That idea became Pathfinder. Isaacson references the Intel/GI cable modem, as well as Zenith. Pathfinder was launched in October of that year and closed in April 1999, a wholly failed experiment. I remember the high hopes that Isaacson had for Pathfinder when I met him in 1994, but the various divisions of Time Warner did not want to provide their content to Pathfinder. It had already been working with online services like AOL.

Still, the cable companies dreamt of controlling content on the Internet, just as they had with cable television. Time Warner, MediaOne, and others created RoadRunner, while TCI, Comcast, Cox, and Rogers created @Home, both functioning as portals with landing sites and email. Time Warner Cable even charged AOL for access. Thankfully, their systems opened up for a number of reasons, including monopoly concerns by US and European regulators raised when AOL and Time Warner were merging. Cable companies were also concerned that they would become what is known as a "common carrier," facing additional regulations. As it turned out, broadband Internet would be totally open; cable companies would have to make money just by providing a high-speed connection, a concept that they first resisted but would discover was a gold mine. More than half of their revenue and profits today come from broadband, though ongoing concerns about open access persist.

Many people, as it turns out, have claimed to be the "creator," "inventor," or "father" of the cable modem. While I certainly have described myself as one of the leaders in creating residential broadband, I have never claimed to be its father. The way memory works is a funny thing; it seems we remember our achievements better than our failures. We also tend to forget or minimize the achievements of others. I have done my best to try to be as accurate as possible. I have devoted great effort in writing this book to research, even though I lived much of this history. Nevertheless, I've most likely made mistakes. I can only apologize in advance for that. Some people may have left out others' accomplishments out of their records on purpose, I'm afraid. With respect to cable modems in particular, the question of "invention" also depends on one's definition of a cable modem. A cable modem, in my view, was something that could coexist on cable's existing bandwidth allocation and would use the cable plant's very limited upstream capabilities.

Moreover, the cable modem was but one piece of the residential broadband puzzle. Many other missing elements were needed to make the network feasible: some way to connect the cable headends and then to access the Internet, software to manage the sprawling network, and applications either created or modified to take advantage of this new high-speed network. My task, as I saw it, was to pull together all of these pieces—and on a global scale. Intel, as it turns out, was a perfect platform for doing this. Even if I did not always have the full backing of the rest of Intel's management, I was given the freedom I needed to accomplish my task—and I did.

ADSL: BROADBAND BY THE TELEPHONE COMPANY

I wanted the regional telephone companies to provide high-speed connectivity for homes, which I thought would push the cable companies to become more aggressive in rolling out broadband service. Some inside Intel also argued that the phone companies could deploy their technology faster than the cable companies could. Besides, many European and other countries did not use cable television. Furthermore, the phone companies had already developed ADSL (asymmetric digital subscriber line) in a fruitless attempt to enter the television distribution business.

We put together a team to work on the technology of using phone lines for broadband under the leadership of Ali Sarabi and Kevin Kahn at the Intel Architecture Labs. I hired people into Intel Capital to explicitly invest in start-ups that would make the critical components needed. We got system companies to produce PC ADSL add-in boards. We worked with software companies to make sure operating systems and applications would support ADSL. Pulling together many companies, Intel formed the Universal ADSL Working Group to develop a standard for ADSL.

Once again, I was on the road meeting with executives, this time, telephone company executives like Ray Smith of Bell Atlantic. They were certainly a different breed than the "cowboys" that ran the cable companies. The spacious executives' suites reminded me of those at

Ford Motor Company, Wall Street, and in some of the government offices. Later, when cable company TCI and Bell Atlantic announced their plans to merge, I knew it would never work.

We set up trials with a number of phone companies all around the world. By 1998, phone companies were rolling out service to consumers. They were indeed able to move faster than cable companies, because their infrastructure enabled two-way communication from the start. But they were uncompetitive with cable in terms of performance.

Unlike traditional telephone modems, DSL uses frequencies above those used by the human voice. The sound of the ADSL modem could nevertheless be heard on the line. That meant that a filter had to be installed between a voice telephone and its connection (or a single filter could be installed after a newly installed split). Later, led by Intel, a variant of ADSL called G.lite became standard. It did not require a filter and could therefore be easily installed by the customer without a telephone technician.

Initial ADSL speeds were low, especially compared to today's, around 1 or 2 megabits per second. As with ADSL, the upstream was much more limited in speed, but this did not matter much because of the asymmetric nature of use. A megabit down was considered good at the time, since, unlike cable broadband connections, each home had its own telephone loop and therefore its own ADSL bandwidth.

The US market wound up split two-thirds cable and one-third DSL. Eventually, telephone companies would use fiber optics direct to the home to significantly improve their performance, in many cases making their services competitive with cable for both broadband and television.

COVAD COMMUNICATIONS

In 1996, a member of my staff, Chuck Haas, had come originally from the Intel sales force, dealing with AT&T. It was his responsibility to help me get the telecom industry's broadband efforts going.

Eventually, Chuck Haas; Druv Khanna, an Intel regulatory attorney; and Chuck McMinn, a consultant, began exploring founding a company to deliver high-speed Internet by DSL through lines leased from the local telephone companies. Regulatory changes required local phone companies to allow third-party companies—known as competitive local exchange carriers (CLECs)—to offer competing services using the telephone company's infrastructure. Chuck Haas and his associates believed that the telcos would be slow to deploy broadband and would be uncompetitive in their pricing and quality of service. They were right.

Haas told me about the idea to create a company. I encouraged him and said that Intel would be a founding investor. I knew the team, and I also thought that they could move first and perhaps cause the major telephone companies to roll out ADSL more aggressively. I committed $2 million and said I would try to help them raise the rest. I approached one of the top partners at Warburg Pincus, Bill Janeway, who I knew well from the Maxis board, where we both served. Warburg, very excited about the opportunity, put in $8 million. One of the founders had a relationship with a VC firm called Crosspoint, which invested, I believe, $1.5 million. So Covad Communications was born and had quite a run for a while.

In 1998, Covad raised an additional $152 million in debt and equity. The company subsequently went public, and Warburg really made out; it cleared about $1 billion on its $7 million investment. At that point, it was the best investment that Warburg had ever made.

Covad was initially focused on the enterprise market. Its first two customers were Intel and Cisco, who offered their employees Covad's ADSL service at home in areas that they served. We then invested in a few other similar companies, such as NorthPoint Communications. A June 1999 article in the *Washington Times* quotes Les Vadasz saying that the Intel investment in Covad was then worth $162 million, which certainly became worth a lot more once Covad had its IPO in 2000. By the time Intel sold its shares, however, it received just $38 million for its $2 million investment. Covad was definitely a strategic success, as it moved before anyone else to provide residential broadband over the telephone copper infrastructure, getting DSL going and providing much-needed competition for the cable companies.

In July 1998, Covad's board brought in a CEO, Bob Knowling, a very experienced executive who had served in senior roles at many important telcos. His last position had been at US West. I think it is safe to say that Bob was probably not the right person to lead Covad at that time. In his 2011 book, *You Can Get There from Here: My Journey from Struggle to Success*, Bob uses Covad as an example of the need to address underlying conflicts in an organization most especially when things seem to be going well, a very important lesson.

At one point, Covad's market capitalization exceeded $10 billion dollars. Unfortunately, Covad and the other CLECs supplying residential broadband did not fare well in the 2000 Internet equity meltdown.

Intel made many other investments in DSL Internet companies, including NorthPoint, a competitor to Covad, and various semiconductor companies.

SATELLITE BROADBAND

I was also interested in the potential for satellite to provide residential broadband in areas that could not be served by cable or telephone. Though we had many discussions with DirecTV and EchoStar, nothing came from these discussions. We also explored working with Loral Space & Communications. Satellites like those used by DirecTV and EchoStar were called geostationary, as the satellite's particular distance from the Earth allowed it to travel at exactly the same speed as the Earth's rotation. A receiver or transmitter could therefore just be pointed once at the satellite. By contrast, Loral and others, like Motorola with its Iridium project, planned to use satellites in low Earth orbit (LEO), with many satellites rotating around the Earth much more quickly than the Earth's rotation. Receivers and transmitters would be handed off between satellites as they went by. In a sense, the idea was like cellular phone service, but with the receiver or transmitter standing still while the towers move.

LEO has several big advantages for telecommunications. First, latency—the time it takes to transfer information from the satellite to the Earth—is far shorter than from geostationary orbit. Latency is very important for real-time communications, such as voice. Second, the amount of bandwidth is much higher, because the area served by each satellite is smaller. Though some of the LEO projects launched, none of the many attempts earned a commercial return on the massive investment required. There is a new round of interest in LEO telecommunications as I write this book. For instance, SpaceX, owned by Elon Musk, has begun a LEO project called Starlink.

We did put together a major program in Europe with Société Européenne des Satellite (SES), a leading European space company based in Luxembourg. Working together in 1997, we launched European Satellite Multimedia Services S.A. (ESM), a company chartered to enable satellite delivery of multimedia content directly to PCs in Europe. The general concept was that large amounts of data, such as multimedia or application downloads, could be transferred via satellite to a PC. This venture had little impact on the market.

PCCW

In 1996, my friend John Evans introduced me to Richard Li, the second son of Li Kai Shing, one of the most powerful and wealthy men in China. Richard had developed what became Star TV. The original idea came from Michael Johnson. Michael, an American film producer living in South Africa, read about a failed communication satellite, Westar, originally owned by Hughes Aircraft Company. The insurance company, Lloyd's, ended up owning Westar. Michael approached them and was, surprisingly, able to get the option to buy the bird. His vision was using it to provide television to 3.5 billion unserved people in Asia. But he did not have the resources, so he approached Li Kai Shing. Li thought this would be a great first project for his second son, Richard. It was.

They were successful in acquiring the satellite. The service was called Star TV Asia. A year or so later, Rupert Murdoch invested $525 million for two-thirds of the company, and in 1995, he bought the remainder for $299 million, making Richard very rich in his own right. Star TV Asia would eventually reach close to one billion people.

Using his share of the profits, Richard started the Pacific Century Group. He did not want to join the family business, perhaps concerned that he would be overshadowed by Victor Li, his older brother and heir apparent to Li Kai Shing's empire.

Richard was subject to a noncompete in the television business and became interested in the Internet. I helped Richard learn about the Internet as a favor to John, including advising him not to buy Web TV (which Microsoft ended up doing).

Richard would have met John Evans during the sale of Star TV. John worked for Murdoch, at one time running all his magazines, including the *Village Voice*. I met John in early 1993 at my first TED conference. He was one of the most extraordinary people I have ever met. For instance, John understood the importance of developing digital map applications and devices. In 1989, News Corp acquired the pioneer in this field, Etak, which had been started in 1984 by Stan Honey and financed by Nolan Bushnell, the founder of Atari. John saw that maps could be transformed into a medium for advertising. MapQuest, the first online map service, was started much later, in 1996.

A bit later that year, Richard reached out. He wanted me to meet with two of his key people by phone about a potential business venture between PCG and Intel. I agreed, of course. I spoke with Michael Johnson and George Chan, the very people who had created Star TV. They proposed that Richard and Intel create a venture to provide Internet capabilities by satellite to rural China. I was intrigued. At the time, little of the world could be reached by wires. The United States had less than 5 percent of the world's population.

Michael and George flew over to meet with me, members of my team, and members of the Intel Architecture Labs. We eventually came to terms and set up a team in Palo Alto as part of a joint venture called the Pacific Convergence Corporation. The former head of Intel Japan, Bill Howe, would be CEO. Michael and George moved to Silicon Valley to join the new venture.

In the mid-1990s, China had very few television stations, and they were all terrestrial. Only a small number of very privileged Chinese citizens had access to the Internet. Since it was hard to receive television signals, cities had headends, sometimes each serving millions of people. From the headends, television signals were distributed to homes. Unlike in the United States, signals were not carried by coax cable, for a funny reason. The valuable copper would have been stolen from the ground. As a result, the service providers ran fiber-optic cable to homes instead.

Using a satellite that could cover China and other Asian countries, our idea was to download and cache Internet data, just like Akamai or other content distribution networks do these days. We would have kept commonly accessed data up to date. Homes could connect by broadband over the fiber and browse. A number of portals would have covered various interests and offered original programming. The system could handle email in almost real time, because it used comparatively little data, especially in those days. It could also download movies and stream them. We had explored similar concepts with other satellite companies, most notably in Europe.

It was all very interesting, though it seems like a dumb idea now. Internet access in Asia developed very differently than we thought it would, and it is unlikely that the Chinese government would have gone along with our ideas. However, some of the concepts allowed PCCW to launch Now.com, which you will read about in Chapter 18.

15. GROWING PAINS

The opportunity to make the personal computer the primary interactive device in the home required not only that cable and telco companies be convinced to pursue this but also that they were capable of the execution. I had many concerns about this.

THE CONCEPT OF @HOME IS BORN

Even as the cable companies moved toward offering residential broadband, the more I worked with them, the more I grew concerned they did not have the technical and marketing expertise to successfully launch a broadband service for PCs.

For that reason, I devised an idea that eventually became a company called @Home. A new company, I imagined, could handle all the novel issues of cable-based residential broadband: selling and servicing cable modems, providing back-end servers, and striking content deals. General Instruments and Intel would provide the initial technical support and financing. Cable companies would offload their potential residential broadband businesses to this company and take a share of subscribers' fees. They would get ownership in the new company based on the number of subscribers using their infrastructure. The cable companies would also be expected to upgrade their cable plants.

I mistakenly included General Instruments as a partner because we had partnered in developing the cable modem and because, as the single largest provider of equipment to the cable operators, they had very strong relationships with the cable operators at every level.

In the fall of 1993, Dan Akerson (GI's CEO) and I visited all the major cable operators with our pitch, traveling on GI's corporate jet all over North America. Though the cable executives seemed interested, nothing happened. Instead, they were preoccupied with telephony over cable to compete in the voice business with the phone companies. Akerson also lacked strong personal relationships with the cable executives. I still had not yet understood that the cable operators wanted to reduce GI's market power, not to increase it. I remain surprised that I missed this at the time and that no one in the cable companies told me.

Frustrated by my lack of progress with the cable companies, I worried that my vision of high-speed access for home computers would never materialize. Then, at the 1994 Western Cable Show, I ran into venture capital legends John Doerr and Vinod Khosla of Kleiner Perkins (KP), who I knew pretty well, having breakfast in a small coffee shop at the Anaheim Convention Center. I was a bit surprised to find them there; Silicon Valley VCs were not normally interested in the cable industry. Looking back, it is clear to me that Bruce Ravenel at TCI must have played a role in their attendance. Like Doerr, Ravenel had once worked at Intel. They were friends. Ravenel was well aware of our plans, as he had attended the February 1994 meeting at Intel where we had presented our intentions to proceed from the initial Viacom and Comcast trials and develop cable-based broadband. Ravenel must have told Doerr about various opportunities the cable industry was considering.

I asked Doerr and Khosla if I could join them. I told them about my idea and said they should consider implementing it, because I could not seem to get it done. I thought it would be very important to create something like @Home if we were to have a successful launch of cable broadband. If I couldn't do it, maybe they could. @Home would go public less than two and a half years later with a market capitalization of $1.2 billion. It would reach a maximum value, after an ill-fated merger with web portal Excite, of $35 billion in April 1999. On October 1, 2001, the combined entity, Excite@Home, would file for bankruptcy.

In early January 1995, Matthew Cowan, a key member of my group at Intel, and I met with KP to discuss my vision for the development of high-speed broadband over cable. On their side were John Doerr, Vinod Khosla, Bernie Lacroute, and Floyd Kvamme. In addition to laying out the concept for @Home in more detail, I pitched KP on investing in Hybrid Networks. At that point, Intel had lent Hybrid $1 million and was looking for a VC to lead the actual round. I believe that KP was interested, but Accel and Sequoia gave Hybrid better terms. This must have been just before Will Hearst joined KP, since he was not at this meeting.

Doerr describes just such a meeting in which the concept was drawn on his whiteboard, but he leaves my role out. He also says that he got the idea while at the 1994 Western Cable Show after he saw a box in the Motorola booth being used to provide telephony over cable. That could be, but most likely, he confused that with the cable modem Intel was showing at the GI booth. Will Hearst—William Randolph Hearst III, grandson of the newspaper baron, and the first CEO of @ Home—also describes this meeting in a 1997 *New York Times* article called "The Seeds That Grew a Cable Network," though he states erroneously that I had already started a company to execute the idea.

Doerr moved the concept forward, building on his relationship with Ravenel at TCI. John Malone, TCI's CEO, was the most powerful person in the cable industry at that time, even though TCI had the worst US cable plant. John Malone's extremely interesting story is documented in the 2002 book *Cable Cowboy* by Mark Robichaux.

Malone is a financial genius, at least with respect to his own interests, and he decided that KP and TCI should form @Home, even though TCI had done nothing yet to even test cable modems. Malone saw this opportunity as a way to get leverage on his fellow cable owners, as well as to profit from them. Once the company was formed, many of the other cable companies later joined as investors, partners, and board members at significantly stepped-up valuations, including Comcast, now by far the leading US cable operator (but one of the smaller operators at the time).

The company that launched differed in many significant ways from my original concept. @Home had no direct relationship with the end user. The local cable companies sold the broadband services, including installation and ongoing technical service. @Home was primarily the back-end network. The revenue was split 65 percent to the cable companies and 35 percent to @Home. Once the cable companies became sophisticated about how the Internet works, they realized they could implement their own back-end networks.

Will Hearst joined KP as a partner in January 1995. I suspect that discussions about his joining had been ongoing for a while. I knew Will when he was still the editor and publisher of the *San Francisco Examiner*. Doerr decided to place Will as CEO of the new @Home venture. Though Will had no technology background, he is a good guy and had some business experience, with a strong sense of technology's potential to change the publishing and communications industries. Doerr and Hearst understood that @Home would need a very strong technical base and looked for someone to lead it. They found that in the person of Milo Medin, thirty-two years old and working on computer networks at NASA's Ames Research Center. With Milo as CTO, @Home was formed in March 1995 with two founding investors, KP and TCI, that together invested $10 million for one million shares of preferred stock. (The company also had founder shares and common stock.) TCI held 75 percent of the preferred shares, and KP held the other 25 percent, giving Malone control of the company.

Mike Ozburn, who worked for General Instruments, met Milo soon after @Home started. Its offices were still bare, without furniture. Boxes were piled everywhere. Milo began the meeting by saying he wanted to give Mike his "vendor test" to see if it was worth Milo's time to meet with him. He drew the open systems interconnection model (OSI model) on the whiteboard, without labels. Milo asked him to label each layer.

Once Mike passed the test, Milo told him about @Home's requirements for a cable modem. The modem had to be external and use cable upstream for the return path. The price had to be under $100. Mike told Milo that almost no cable systems had an adequate upstream; "we know, because GI built them," he explained.

As soon as Will Hearst was announced as CEO, he came over to Intel to meet with me. Since @Home was my idea in the first place, he wanted to learn everything I already knew. He also explained that one of his tasks was to find a real CEO to run the business, which would take him almost a year. He either asked me directly if I was interested in the role or hinted at it. I made it clear that I personally had no interest in being the CEO of this or any other venture. After my earlier bad experience at Franklin Computer, I had decided that the CEO role was not a good fit for me. I also already had a great job at Intel, where I was motivated to work on rolling out residential broadband worldwide; being CEO of @Home would have felt limiting by comparison. Nor could I imagine reporting to a board of John Doerr and a bunch of cable people, not that Doerr would have selected me anyway. He generally had a good sense for who would make a great CEO. That was probably not me, but in the case of @Home, he made a terrible mistake, in my opinion, selecting Tom Jermoluk, COO of Silicon Graphics, as Hearst's successor.

When I learned that KP and TCI were setting up @Home, I asked Doerr if Intel could invest. He declined and blocked us from the first round. That really pissed me off. Intel eventually invested just $1 million later at a much higher valuation with some of the other cable companies. This was not the only time that KP, and especially Doerr, took advantage of Intel. I used to joke that the only time Doerr would call to bring Intel into an investment in KP's portfolio was when the portfolio company was either in trouble or overvalued.

After meeting with Hearst, I met with Milo, @Home's CTO. Though I don't remember many details of our conversations, Milo had clearly thought deeply about the back-end infrastructure, but he did not understand the cable system all that well. He had rightly identified some barriers to the large-scale deployment of residential broadband concerning the efficient distribution of content, but his approach was very expensive and not really required in the early days. In other words, he overengineered the network. Eventually, the problems he was trying to solve were addressed by other means and by other companies, such as Akamai and Inktomi. Milo continues today to work on network expansion at Alphabet, the parent company of Google.

@HOME: STRUCTURALLY FLAWED

Technical not structural issues doomed @Home, however. The company's ownership by the cable companies created inherent conflict on the board, as the cable companies were motivated to transfer as much value as possible to themselves and away from @Home, especially once they sold their ownership in @Home. These problems resulted in part because the cable companies saw themselves as @Home's customers and owners, while @Home saw the cable companies as its "distributors" and cable subscribers as their customer.

The late 1990s brought major consolidation of the cable companies. In 1998, AT&T bought TCI (for $48 billion) and then MediaOne in 1999 (for $62 billion). Comcast acquired AT&T's broadband assets just three years later for about half the price that AT&T had originally paid. By then, @Home had gone bankrupt.

I never developed a relationship with Tom Jermoluk, who seemed very political and removed from the company's operational issues. I did develop a good relationship with CFO Ken Goldman, and we have remained friends since. Intel completed a number of deals with @Home internationally. For instance, we created @Home Benelux to provide broadband capabilities to the Netherlands. That company still exists, rebranded as Ziggo.

@Home went public on July 11, 1997, just two years after its formation and two and a half years after my breakfast with Doerr. It sold 9 million shares at $10.50 each, raising $94.5 million. Prior to the IPO, TCI owned 45 percent of the company. The total market capitalization at the offering was $1.2 billion. KP's share was 5.2 percent. The number of subscribers peaked toward the beginning of 2001 at three million.

Prior to the IPO, Doerr tried to persuade AOL to invest in @Home to become the service's landing page. This would have been a smart move for AOL, but they had not yet realized the impact that broadband and the Internet would have on their business.

TWO ROCKS TIED TOGETHER, TRYING TO FLOAT

On January 19, 1999, @Home merged with an Internet portal, Excite, valued at $6.7 billion. The combined company, Excite@Home, saw its stock price rise as high as $128.34, valuing the company at $35 billion. Both companies had flawed business models. For instance, @Home's agreement with the cable companies expired in 2002. I would guess that when @Home was first structured, this did not raise concerns that the cable companies would leave the arrangement because they owned 85 percent of @Home. The merger with Excite, however, reduced their ownership to about 50 percent. Malone was no longer able to dictate to the board once he sold TCI to AT&T. AT&T bought out Comcast's and Cox Communications' stakes in Excite@Home for $1.6 billion each, twice the market price. The transactions were actually more complicated than is worth explaining here, extending over several years of negotiations and renegotiations. The bottom line is that AT&T ended up paying more than $4 billion for control of Excite@Home, screwing itself to the benefit of Comcast and Cox.

As for Excite, the portal business at the time had questionable merit, since users seemed wholly unreceptive to the "banner" advertising common on the web at the time. (Advertising based on search, perfected by Google, is much more appealing to users and makes far more money.) Moreover, because of the nature of the Internet and the experience of using a web browser, there was no particular reason for cable customers to keep Excite as their landing page. They could very easily change their portals to Yahoo or AOL. Had Excite developed a website that took real advantage of the broadband connection, there might have been a reason for cable customers to favor Excite. On the other hand, there were not that many cable broadband customers, which discouraged the Excite business from making such investments.

The merger between Excite and @Home always seemed very strange to me, perhaps not even completely kosher. KP was behind the merger and had stakes in both companies. It reminded me of another KP deal between GO Corporation and EO, Inc., about which I also have some questions. Besides the shared KP ownership and the fact that the two companies were located next to each other in Redwood City, California, little made sense to me about the merger. Perhaps it was thought that Excite could build a portal that took full advantage of the high-speed cable broadband network, but that is not what they did. Instead, Excite@Home bought an electronic greeting card company, Blue Mountain, for $780 million. The founder of that company, Jared Polis, is today the governor of Colorado. In September 2001, Hallmark purchased Blue Mountain for $35 million.

In the early days of @Home, a lot of money was made. Early investors won big, including the cable companies, KP, and the founding employees. Intel also made a substantial profit on its modest investment. Those who bought the company's stock during the Internet feeding frenzy of 1999 and early 2000 lost big, but AT&T lost the most. After filing for bankruptcy in October 2001, Excite@Home ceased operations in the spring of 2002. This very same year, Comcast bought AT&T's broadband division, comprising the former TCI and MediaOne. From that point on, Comcast would be the leading US cable company.

From Intel's point of view, @Home was successful. We achieved our strategic objectives; the cable industry fully embraced the PC and deployed high-speed residential broadband. The cable companies involved got the capital they needed to deploy their broadband networks. As AT&T found out the hard way, not everyone can be a winner.

ROADRUNNER: HOW MICROSOFT OUTMANEUVERED ORACLE AND INTEL

Toward the end of 1997, I was contacted by Dave Roux, the executive vice president of business development at Oracle, about an opportunity to invest in RoadRunner, the broadband ISP for Time Warner Cable and MediaOne Cable. Though RoadRunner functioned similarly to @Home, there were several important differences. RoadRunner was the brand that was promoted, which had been part of my original @Home concept. Although it, too, was owned by a combination of cable companies—Time Warner, MediaOne, and Newhouse Media— its board was not nearly as dysfunctional as @Home's. The management of RoadRunner came from the cable industry. They had worked for the companies that owned RoadRunner and had strong and positive relationships with those companies, while the executive team of @Home came from the computer industry and had difficulty relating to cable company employees.

Roux explained that the RoadRunner team wanted computer-industry expertise and endorsements. How the initial discussions between Oracle and RoadRunner began, I cannot recall. Oracle was very interested in getting a toehold with the cable industry. Roux, who later founded the very successful technology buyout fund Silverlake asked me if Intel would like to partner with them in investing and working with RoadRunner. I readily agreed, especially since we had been squeezed out of any major investment in @Home by John Doerr and John Malone. And importantly, I wanted to see RoadRunner be successful in deploying residential broadband.

The RoadRunner team was led by Glenn Britt, who would become the CEO of Time Warner Cable in 2006. Glenn was a wonderful man, and we developed a friendship. Sadly, he died too early, in 2014.

Oracle and Intel formed a team to negotiate with RoadRunner. We met repeatedly in very productive sessions, working not only on the investment terms but also on various technologies that would be developed. We wanted to make sure RoadRunner would be a great success, primarily for strategic reasons, though, of course, we eventually wanted to make a great financial return. We agreed that Oracle and Intel would each invest $125 million for 10 percent of the company, or 20 percent in total.

For an investment this large, I had to get the approval of Intel's board, in addition to Andy and Les. Les was, immediately, very supportive. Andy's approval came a lot easier than I anticipated. By this time, in early 1998, Andy understood that I was correct in my belief, first expressed five years earlier, that the cable industry would be the leader in high-speed residential broadband, at least in the United States.

Once we reached an agreement with the RoadRunner team, we planned an announcement for a Monday morning in New York City. Oracle, Intel, and RoadRunner would first meet to sign the formal agreements, and then we would meet with the press. It was Sunday morning when Glenn Britt reached me by phone. I was already in New York. What he told me left me stunned. Evidently, Bill Gates had learned about the deal we had put together. He went nuts, very upset that Oracle was involved. Somehow, he reached Gerald Levin, Time Warner's CEO. Whatever Oracle and Intel were offering, Gates told Levin he would double it. Gates didn't even know the amount. Levin decided that he had to take Microsoft up on their last-minute offer. Frankly, this was incredibly shortsighted, because Microsoft would do little to nothing to help RoadRunner. It was just about the money. Levin could also have tried to negotiate with the Oracle and Intel team. Knowing Bill at that time, he probably used a carrot-and-stick strategy, but I don't know what he used as the stick. But Levin was not very astute when it came to these kinds of deals, as he later showed in Time Warner's merger with AOL. Bill once again demonstrated his aggressive, competitive behavior.

Glenn was very unhappy about Levin's decision and apologized but said he had no ability to change it. I was, of course, extremely upset, even embarrassed. I had taken this deal all the way to Intel's board! Then again, Les, Andy, and the rest of the board all knew Gates and were not surprised at his assault.

Tuesday, back in my office, one of Microsoft's top executives called me. He is no longer at Microsoft, and he currently plays a big role in the cable industry. He said, "Intel can join our team, in the same role you would have had. It's not that we didn't want Intel in the deal. We just didn't want Oracle." I told him that we were partners with Oracle. We were not going to turn our backs on them, I said. Then he said something very remarkable. "We want to get the deal done fast. Can you please send me the agreement that you had already negotiated with Time Warner? We'll just change the amounts and substitute Microsoft for Oracle." I said, "I can't believe you just asked me that." Then I hung up on him.

Time Warner required that Microsoft find another technology partner. They didn't want Microsoft to have more than 10 percent of RoadRunner. That's probably why Microsoft offered Intel a chance to remain in the deal. When we rejected it, they turned to Compaq. Craig Mundie contacted Bob Stearns, vice president of Compaq. Bob likely did not know that Gates had wrestled the deal away from Oracle and Intel. Compaq agreed to join, but why they did so is not clear to me.

While Oracle and Intel had expected to work closely with RoadRunner, Microsoft and Compaq were very hands-off. Later, because of complexities resulting from the merger with AOL, Time Warner bought the RoadRunner shares from Microsoft and Compaq for $570 million, about a 34 percent gain on their original investment made just eighteen months prior.

RoadRunner has continued in some form ever since. It became part of Time Warner Cable, which spun out of Time Warner in 2009. Eventually, Time Warner Cable was purchased by Charter Communications, which was, strangely, controlled by Microsoft co-founder Paul Allen until his death in 2018.

It probably didn't matter to Intel, in the end, that we did not invest in RoadRunner. Microsoft's and Compaq's investments only helped us toward our strategic objective to grow the market for home computers.

PART IV

FLYING SOLO (1996–2002)

*Surviving prostate cancer and
leaving Intel to be on my own.*

16. CANCER SURVIVOR

FOLLOWING IN ANDY'S FOOTSTEPS

In May 1996, Andy Grove wrote a very influential article in *Fortune* on prostate cancer. He had been diagnosed and treated for prostate cancer the year before. Andy did absolutely everything with great conviction, focus, and discipline. The article described his journey from when he learned that he had prostate cancer to his decision about what kind of treatment he would have. Like so many, I read it with great interest, not realizing that I would soon join this rather nonexclusive club.

A month later, at a routine physical, my doctor noted that my PSA (prostate-specific antigen), an indicator of potential cancer, had increased to 4.3 since my last physical one year earlier, slightly elevated from normal for a man of my age (fifty-one). She suggested that I consult a urologist, which I did, following, in part, what Andy had written in *Fortune*. The urologist recommended a biopsy to determine if I had prostate cancer. This was the beginning of a very miserable four-year journey for which, perhaps unexpectedly, I was actually grateful in the end.

The urologist told me that the odds prostate cancer would be found in the biopsy was below 25 percent. I subsequently learned that the chances of finding at least microscopic prostate cancer correlate highly with one's age, so this was a remarkable claim; the true chances were probably closer to 50 percent. I remember discussing the situation with Andy at the 1996 Allen & Company conference in Sun Valley.

He encouraged me to have the test. I had the biopsy soon after. I can attest it was not a lot of fun.

IT'S NOT A GOOD SIGN WHEN YOUR DOCTOR ASKS YOU TO COME AT THE END OF THE DAY

A week later, I came back to speak to the urologist about the results. His assistant made the appointment for the end of the day. I sat there in his waiting room until I was the last patient remaining. I knew this meant the news would not be good. Telling someone they have potentially deadly cancer must not be an easy task, though it was also not an unusual task for a urologist. One in nine men will be diagnosed with prostate cancer, the second-most prevalent cancer in men (after lung cancer). For most men who are diagnosed with prostate cancer, however, it grows very slowly; as a result, many men will die with, but not from, this cancer.

The urologist did indeed tell me that I had drawn an unlucky ticket: I had prostate cancer. My Gleason score was six, which was not a very high score under the circumstances. The Gleason score is a way of measuring the cancer's aggressiveness, with a maximum score of ten. He strongly recommended that I treat the prostate cancer, particularly because of my young age. At fifty-one, not only did I have a long life expectancy, but also prostate cancer is typically more aggressive in younger men. I did enjoy the side benefit of being termed a "young man" once again, however.

I asked him if I could wait six months while I investigated the best approach to take. He could not recommend waiting, he said. I used the following tactic to understand the actual urgency. I asked, "Do you think the chances I'll die from prostate cancer will increase by ten percent if I wait six months?" He said no. "Would they increase by five percent?" I asked again. "No," he replied again. Finally, I asked, "One percent?"

"I don't think so," he concluded. I told him I would take my time to figure things out. I realized, after reading what Andy wrote and talking to him, that urologists are committed to doing surgery for prostate cancer, which they consider the gold standard. The idea was that if you could remove the prostate while the cancer was still contained within the prostate capsule, you would then be cured of prostate cancer. As it turns out, this reasoning was fallacious in many respects. Urologists are basically surgeons trained to cut; they don't call them "sword men" for nothing.

Twenty-five years ago, when I was deciding what to do, the concept of "active surveillance" was not really an option. Now, we understand that treatment can be avoided or at least delayed with adequate surveillance, which means having a biopsy every year or so. Many men who cannot tolerate the ambiguity of that prefer to have their prostate removed, which they think means the cancer is gone for good. Unfortunately, that is not always the case. Deciding what to do is very complex and stressful, although most patients probably just do whatever their first doctor tells them to do. There was a lot more overtreatment back then. (It is better now.)

Walking out of the urologist's office, I was in shock. I broke the news to my family; my former wife, Arianne; and then to my girlfriend, Harriet. My task was to make sure they were calm and did not think I was about to die. At that point, I figured I had a greater than 90 percent chance of survival.

Soon after, I had a consultation with a radiologist, Dr. Gordon Ray. He explained how prostate cancer is treated with external beam radiation. He reviewed the side effects of radiation and compared them to surgery. The research I did had already convinced me that I would not do surgery. One downside of having your prostate surgically removed is that, very often, you are rendered impotent. It didn't seem like this was the right time of my life to experience impotency. The results for radiation were much better with respect to morbidity, including this type.

I spoke with Andy many times during this period. He was always there for me, although he was extremely opinionated, as usual. Andy had chosen a new treatment called internal seed radiation, or brachytherapy. He became a big advocate. The leading centers for brachytherapy were in Seattle, where he had been treated. With his help, I arranged to go there. I met with a doctor, a wonderful man with the unfortunate name of Dr. Grimm. Dr. Peter Grimm sadly died young of a heart attack in early 2016.

In addition to just doing brachytherapy, it was also possible to receive a kind of hybrid treatment, a combination of internal and external beam radiation. This combination had the benefit that any cancer cells outside of the capsule might be radiated. I wasn't sure what to do. Dr. Ray, my radiologist at the Palo Alto Medical Foundation, advocated for the combination. Dr. Grimm disagreed; he thought brachytherapy was sufficient. One frequently gets conflicting advice from doctors, which leaves it to the patient to try to sort out the best approach. One night I had a nightmare. In it, I died because I did not do the combination. So that is what I did. If I could go back, I probably would have just watched and waited for a while.

ZAPPED

In November 1996, I went to Seattle and checked into a hotel. I decided I would go for treatment on my own. I arrived early the next morning. I was given a strong local anesthetic in my pelvic area, similar to what is sometimes given to women during childbirth. They brought me into a small operating area. Three people were there: Dr. Grimm, a resident he was training, and a technician.

I watched the entire procedure on a screen and felt no pain as they inserted sixty tiny radioactive balls into my prostate. They are still there, though the radiation dissipates after some number of months from the insertion. I found the procedure interesting, but I was surprised to learn that I would be treated not by Dr. Grimm but by the doctor in training. I thought they should have informed me of this in advance. After the procedure, I rested until the feeling in my lower half returned. I then left and decided to reward myself with a large chocolate-and-nuts candy bar, even though I was on a no-fat diet.

The next day, I went back in. They said everything was OK, took some X-rays, and I returned to Palo Alto. I went to work right away, even though I had to do ten weeks of external radiation. Dr. Ray put three small tattoos on my body so that the laser could help them focus the radiation machine exactly on the right spot. Five days a week, on my way to work, I went to the Palo Alto Medical Foundation, pulled down my pants, laid on a bed, and got radiated. Afterward, I drove to the gym and worked out for an hour. I decided that I would end the radiation treatments stronger than when I started, and I did. I worked every afternoon as if nothing at all was going on.

Soon after I had the radioactive seeds injected, I learned something I had not expected. Nobody had explained to me that I would no longer be able to ejaculate. My prostate was effectively gone, and its job had been to provide seminal fluid. I could still have sex and could still orgasm, but the orgasm was different. I was also upset once I realized I could no longer have children. Though I was already fifty-one years old, had three children, and had no interest in having more, I didn't like the idea that I couldn't. That somehow changed my sense of self. This could have been easily prevented by taking semen samples from me and freezing them. I'm sure I would never have used them, but the idea that I could have would have been important to me.

After a week, I started to get side effects from the radiation. Though I had been warned about this, the experience was far worse than I'd imagined. My bladder was very inflamed. I constantly had to use the toilet. It was really scary. I remember being afraid of getting into the car to drive to a meeting, and flying on airplanes was terrifying. Fortunately, Intel agreed that I could fly first class, so at least I had a better shot at getting to the toilet before I wet my pants. This went on for about three months and caused me much anxiety. Others I know who were treated with radiation have related similar experiences. After about six months, the radiation from the seeds died away, and my bladder problems were gone.

Andy, along with many others, believed that an extremely low-fat diet was particularly important with respect to survival. Though I tried to follow this diet, sometimes I didn't. If I was having a meal in the Intel cafeteria and Andy was there, he would often come by to check on what I was eating. Once he caught me eating cottage cheese and began to scold me. Others overheard him. I know he did it out of concern, but it was very embarrassing. From then on, I took my food into a conference room.

Every month, I had another PSA test to see if the radiation was being effective. Prostate cancer cells die as a result of radiation, so the amount of PSA manufactured would diminish. With surgery, PSA is supposed to go to zero quickly, because you don't have a prostate. If that is not the result, it means real trouble. With radiation, PSA usually falls below 0.5 percent, but this can take about a year or so.

After six months, my PSA had not come down. Dr. Ray was a bit surprised, and so began one of the most frightening periods of my life. Dr. Ray said, "Well, at least it's not going up." A few months later, it started going up. That was a really bad sign. He said, "Well, at least it's not going up quickly." It started going up quickly. Then I got really scared. I was taking PSA tests every few weeks, but it took quite some time in those days, unfortunately, to get the results. I would be in terror waiting for Dr. Ray to call me and tell me what was going on.

I had been thinking, around this time, about leaving Intel, and I explored the possibility of starting a leveraged buyout company with my friend, Dave Roux, the executive vice president of business development at Oracle. As my PSA rose, I no longer wanted to commit to starting a fund. I told Dave I could not be his partner. Based on many of the ideas he and I had explored, he founded Silverlake, which became a very successful private equity firm with $43 billion under management. I had just given up my chance to become a billionaire.

MIKE MILKEN TO THE RESCUE

Andy Grove had developed a relationship with Michael Milken, the famous junk-bond dealer and convict. Milken had started a foundation, CaP CURE, that funded research on prostate cancer. Andy was a supporter. I became involved and eventually joined the board. I met Milken a few times. Andy often had me meet with individuals who were important to him and had shown an interest in learning about the Internet because I was "user friendly," which meant I could explain things well, did not make people uncomfortable, and dressed nicely.

Milken had prostate cancer that had actually spread beyond the prostate, but he was somehow able to keep it in check, particularly, at least in his mind, by controlling his diet. I spent time talking to Milken about my situation, and he helped me set up a schedule of visits to top specialists around the United States. As these doctors had received grants from CaP CURE, they were very responsive.

Michael Milken and me

I took time off from my duties at Intel to manage meetings with the various doctors. My personal assistant compiled binders of test results, which we sent to various institutions. My mind was not really focused any longer on making minority investments in early-stage companies. I thought that perhaps I would have to fight for my life for the third time, following my fights with chronic childhood asthma and the collapse of my left lung in mid-1963. I requested a medical leave of absence from Les, which he granted, and I transferred my staff to Steve Nachtsheim.

I scheduled meetings with the Memorial Sloan Kettering Cancer Center in New York City, Johns Hopkins in Baltimore, MD Anderson in Houston, Cedars-Sinai in Los Angeles, and UCSF in San Francisco. Every single doctor told me something different. One group said my prostate cancer had clearly metastasized; I probably had five to ten years to live. "Enjoy your life," they said. Another group said, "We need to do everything possible to stop it!" They suggested a procedure called "salvage surgery" that would have left me without any real plumbing. The others simply admitted they had no idea.

Of all the prostate cancer doctors I met, I became closest with Dr. David Agus, a brilliant man with many talents who has written several books. David is a rare blend of a caring doctor and an outstanding scientist. While he has a very busy life, he has always had time for me and has encouraged me to continue to be active and challenge myself.

I discussed his first book, *The End of Illness*, with him while he was working on it. David was one of Steve Jobs's doctors at the last stage of his life. Jobs and I gave David conflicting advice on the title of his book. Steve won out, beating me again, one last time. It was amusing to me to see my name next to Steve Jobs in the acknowledgments section of the book, which was published in 2012.

The question remained: Did I have metastasized cancer? David Agus arranged a PET scan of my entire body, done at UCLA. I could not help myself; I was really interested in the technology. They injected me with some kind of radioactive sugar that was supposed to be taken up by any cancer cells. Then we went on a hunt for areas that might "light up." Thankfully, we saw nothing suspicious. I got them to let me operate their computer system and had the amazing experience of flying around inside my own body.

My PSA was now over thirteen. I decided to have a biopsy of my radiated prostate at UCSF, which found cancer cells in my prostate that were still very much alive. I convinced the cancer doctors to let me sit with the pathologist and review my slides. He showed me my cancer cells. They were very mature and large and were probably throwing off lots of PSA. We agreed that the cells also looked very damaged; they were probably not capable of dividing and reproducing. I got to keep some of the pathology slides. Later, Dr. Ray asked me to present these "findings" to the radiation department at the Palo Alto Medical Foundation, which I did.

Once I accepted that there really was not much to be done, I grew calm. I focused on my health, doing yoga, having massages, and eating well. I took my second sabbatical from Intel. Harriet and I took a trip around the world. We traveled to South Africa, a country I love, as well as Hong Kong, Bali, and London before returning to her home in New York. It would be our last trip together. We had been in an on-again, off-again relationship for seven years. While we both cared a great deal for each other and enjoyed being together, it had become clear to both of us that our relationship would not progress.

I also started raising money for CaP CURE, the nonprofit Mike Milken had started. I reached out to many of the people I had worked with on the development of residential broadband, raising a million dollars from key individuals in the entertainment, communications, and venture capital industries. Their generous response to my requests touched me deeply.

RETURNING HOME TO CALIFORNIA

In 1997, I had decided to build a home in Sonoma wine country, located sixty miles north of San Francisco. Sonoma played a big role in my family dating back to the 1920s. I spent many summers there and thought I wanted to have a home in Sonoma that I, my children, and eventually their children could enjoy. Having looked for such a home for many years, I finally decided to design and build my own. It was a four-year project. I bought a very beautiful property, surrounded by vineyards, in an area that looked like Tuscany. After spending a year learning about architecture and construction, I hired an architect, and we began working on the design. That took eighteen months. The home would be constructed of rammed earth, comprised of 80 percent clay, 10 percent concrete, and 10 percent sand. The walls would be eighteen inches deep. Just before we started construction, which would also last eighteen months, I had a decision to make. At this point, my PSA was rising rapidly, likely indicating that the cancer had metastasized. I had to decide if it made sense to go ahead with such a massive project. Then I realized that I wasn't building a house but a work of art that would outlast me. I was also very sure that I would survive long enough to see it completed.

In April 1999, I returned to San Francisco, the city of my birth and childhood. The home I was building in Sonoma was a country home; I needed a place to live in San Francisco, as well as an office. I saw no purpose in staying any longer at Intel. I wanted to be out on my own and work for myself. Since it appeared likely that I did have metastasized prostate cancer, I thought I might have less than ten years to live.

PSA BOUNCE

Now that I was back living in San Francisco, I did my PSA tests there. In mid-January 2000, I heard from Dr. Peter Carroll, the head of the Urology Department at UCSF. He called to tell me that my PSA score was lower than the last time, the first drop since I was treated in 1996. I started crying. Over the subsequent six months, it continued to drop. Now, twenty years later, it is almost undetectable at 0.2 percent.

The seed procedure used on me was pretty new at the time, so data was scarce. Later, doctors documented a phenomenon called PSA bounce. About a third of patients—disproportionately younger patients—experience a rise in PSA around eighteen months after treatment. (Most men treated for prostate cancer were at least a decade older than I was.)

PSA is a problematic diagnostic test. I don't think I should have been treated so early in the course of my disease. These days, treatment is often delayed for men in a similar situation as mine. Instead of treating the prostate cancer, they are just observed, looking for changes in the PSA and periodically doing biopsies.

Prostate cancer is not a single kind of cancer. Most types will not result in metastatic disease and therefore are not terminal. Unfortunately, we don't know yet which forms will metastasize, but we do have some clues. I think that if most men who are diagnosed with early-stage prostate cancer just lost twenty pounds instead of having their prostate removed or radiated, there would be less mortality overall.

I must admit that while I was not very happy about the experience, I am happy with the result.

Sometimes I joke that cancer saved my life, because I decided I would not continue to work like I had in the past. I wanted to live a more balanced life. I decided that I would make all decisions about the future assuming I only had ten years to live. Ten years is long enough to do something significant—write a book, perhaps, or learn a language—but not so long that you feel you can put things off. It has now been twenty years since I made that decision. It was a good one.

However, at seventy-six, I've shortened my horizon to five years.

17. THE AVRAM MILLER COMPANY

I would have left Intel even if I had not been dealing with prostate cancer. There was nothing more for me to do there. I was mindful that Les would retire in a few years and he needed to groom his successor. I didn't want his job and probably would not have been given it as I was not part of the old boys' network, even having been at Intel for fifteen years.

In 1997, in a breakfast meeting similar to when I first meet Les, I had confided in him that I would not be staying at Intel beyond 1999. I suggested that he should look to bring in someone senior to replace me and who could take over from him eventually when he retired. This is how Steve Nachtsheim, an Intel vice president, ended up at Intel Capital in 1998. However, Steve was not hired for the job and retired in the summer of 2001.

When I returned to Intel at the beginning of 1999 from my medical leave, it would be just for three months. Les asked me to take on a project which we called 10x. The task was to develop a plan to increase Intel Capital by an order of magnitude. Given our size at the time, that would have meant building a portfolio of more than $100 billion. I worked on it for three months and then presented it to Andy and the executive staff. It was a good plan, but there were very few comments or questions, most likely because I was a lame duck. Everyone knew I was leaving the company the next day. Needless to say, Intel did not implement my plan to scale Intel Capital up by ten times. I doubt that it could have been possible. But we could have increased our activity significantly. Perhaps this would have helped Intel navigate the next phases of computing. It appears the opposite has happened. Intel Capital would become more financially driven and play a diminished strategic role once Les retired.

That day would be my last at Intel. Other than Les, no one even said thank you to me for all my efforts or even goodbye, with the only exception being Andy. I was pretty sad but looking forward to the next phase of my improbable journey. For the first time in my professional career, I would be working for myself. I recognized that so much of my success was due to the importance of Intel in the industry as well as the infrastructure it provided me. Now, I would have the opportunity to see what I could accomplish on my own.

NEW IDEAS FROM THE SAME OLD PEOPLE

The appointment of Craig Barrett in 1998 as CEO—followed by the creation of the New Business Group (NBG) under Gerry Parker, who had been running manufacturing—removed any doubt I might have had about leaving Intel. While Craig made many important contributions to Intel, particularly in terms of creating its world-class semiconductor manufacturing capability, he was totally the wrong person to lead Intel through the transformation awaiting it, driven by the Internet. Barrett had been Intel's chief operating officer. It had become an Intel tradition for the COO to become the CEO and the CEO to become board chairman. While Barrett was a poor choice, I am not sure there were better choices available internally, and it would have been very difficult for someone outside of Intel to be successful as CEO. The time to have brought in and groomed a successor to Andy would have been several years earlier.

Once Barrett took over, he changed the company's mission to supplying the "building blocks of the Internet." This would become the company's third phase, after memory components and then microprocessors. Intel was aware that the PC market growth was slowing, and competition from companies like AMD had gotten more aggressive. In my opinion, Intel was at least five years late in making this pivot, which the bursting of the Internet bubble in 2000 made even more difficult. While "building blocks" was a catchy phrase and actually something that Intel partially achieved, it did not recognize that the value in the Internet was owning the "buildings" not in manufacturing "blocks."

Intel's board, including Andy himself, had come to realize that Andy had gotten more and more risk averse. They hoped that Barrett would be a bit of a cowboy, willing to take risks. He was. Craig thought that Intel needed to change and develop new businesses, but he had no idea how to accomplish that. I used to joke that Andy wanted the same old ideas from the same old people, while Craig wanted new ideas from the same old people. Actually, it was true, and it was not really funny. Perhaps some of my criticism stems from not being asked for any input when he created NBG, but then Barrett did not really ask Les either.

Craig created NBG in an attempt to develop new internal businesses. Amazingly, rather than finding a business-oriented executive to lead it, he turned to the head of the manufacturing organization that Craig once lead, Gerry Parker. Gerry had done an outstanding job in manufacturing and was very well respected inside the company, but he knew nothing about business. Gerry then recruited a number of key people from the manufacturing organization, who also knew nothing about business. NBG would start with "Big Green" to provide server hosting and a bunch of small start-ups, which were called "Little Green" or NBI (new business investments). The idea was to function as an internal VC, funding promising internally generated start-up ideas. In a sense, it was modeled on the Intel Development Organization Les had started in the 1980s.

A long-time Intel executive, Mike Aymar, was brought in to run Big Green, which was called Intel Data Services (IDS). Gerry's former financial controller, Rich Delature, would run NBI. While I respected all of them as individuals, I knew they were not the right people to perform their new assignments.

It was not surprising that I had not been considered to run the New Business Group or even the Little Green group, but I was disappointed that no one had even consulted with me on how to set things up. No one at Intel had more direct experience working with early-stage companies than I did. I asked Les if I could spend one day a week helping

Gerry. He, of course, said yes. Gerry, while friendly as always, did not seem interested in my help.

At that time, Intel had a significant stake in CMGI, a company that was really a collection of Internet companies. One of these was NaviSite, which was already well established in the hosting business. I suggested to Gerry that he meet with Dave Wetherell, the CEO of CMGI, to better understand Dave's view of this kind of business. I also suggested that Intel should perhaps buy NaviSite from CMGI. He did meet with Wetherell at my request, but Wetherell told me that Gerry had no interest in learning about CMGI or Navisoft. Gerry had no real clue about starting a system service business, and he put in charge Aymar, who had been at Intel for twenty-five years and never ran a service business. The two of them were ready to invest billions of dollars into something they totally did not understand. Ultimately, Gerry retired from Intel two years later.

*— it's much easier to invest in a start-up
than create one internally*

Gerry did a two-part oral history for the Computer History Museum in May 2016. He briefly discussed this period of his career. Gerry had a remarkable career and had accomplished much. It was a sad way to end what had been a brilliant career. Mike Aymar also left Intel. NBG was closed down. More important than losing a great deal of money, NBG failed at its strategic mission, which was to create new opportunities for the company.

WHAT HAPPENED TO INTEL CAPITAL

Intel Capital is now more than thirty years old. The collapse of the Internet bubble was not kind to the Intel Capital portfolio, and the value of its portfolio fell in 2003 to just $870 million from a high of $13 billion, as the *Wall Street Journal* reported on April 13 of that year.

However, Intel had realized gains of more than $5 billion, and had made more than five hundred investments in its first ten years. It was a great success by anyone's measure. Both Les Vadasz and I are very proud of what we created.

John Miner succeeded Les as president of Intel Capital. He was an Intel vice president and had successfully run a group that sold products designed for Internet servers, taking over NBG after Parker. John left just a few years later, just as Paul Otellini became CEO, which was not a coincidence. Unfortunately for John, he had to spend much of his effort cleaning up the portfolio given the massive meltdown in the technology sector. John was strategic and knew Intel's business well. I think he would have kept Intel Capital pretty much as Les and I had originally intended, focusing on expanding Intel's business, providing strategic insights, and earning an excellent financial return. We believed that corporate venture capital should primarily be a strategic function, especially for companies like Intel. It makes no sense to me that a corporation would have a captive venture arm that operates like normal VCs, even though many companies have done exactly that.

In 2005, Arvind Sodhani, Intel's treasurer, took over as president of Intel Capital. Arvind, I suspected, had wanted that job for some time. He and I had worked closely together in the past. It was not always an easy relationship. Though he was an excellent treasurer, Arvind was not really grounded in Intel's business. He was anything but strategic, in my opinion, focusing more on financial returns. He even changed Intel Capital's compensation scheme, rewarding members of the team, including himself, on the portfolio's financial performance. He expanded Intel Capital's footprint all over the world, driven, too, by financial return. CEO Paul Otellini took a hands-off approach to Intel Capital, it appears. Unfortunately, Paul was not very strategic himself, focusing primarily on the core microprocessor business, or as Andy had termed it, Job 1. He also had a lot on his plate, including laying off more than 10 percent of Intel's workforce. Paul retired early in 2013, replaced by Brian Krzanich. Sadly, Paul died unexpectedly young in 2017, at the age of sixty-six. I liked Paul personally, but he was not a good choice for CEO. He was basically a finance person at heart and not willing to take chances. He totally missed the importance of mobile computing, turning down the opportunity to manufacture the microprocessor for the iPhone in what will go down in the history of technology as one of the greatest strategic mistakes of all time. One that may prove terminal for Intel.

Krzanich, who I did not know, was much worse, it appears. He was terminated in June of 2018, allegedly for having a sexual relationship with an Intel employee. While perhaps true, my guess is that it was just a cover. The board did not want the market to get a sense of the many problems Intel was now facing.

— a board's primary responsibility is to make sure the company has the right CEO

Bob Swan, the CFO, was acting CEO until he was given that role permanently in January 2019. A board's primary responsibility is to make sure the company has the right CEO. For twenty years, from 1998 to 2018, they failed at this task. Intel has seen a steady decline in its importance in the computer industry over the last twenty years. I believe its board of directors bears the primary responsibility for this. Now, in 2021, Intel is no longer the leader in semiconductors, with companies like Nvidia having a higher market capitalization. The semiconductor industry had been reconfigured, with companies such as TSMC and Samsung being foundries for companies like AMD and Apple. Intel, which manufactures and designs its own semiconductors, shockingly fell behind in manufacturing technology. Recently, the Intel board finally took some real action by terminating Swan and putting Pat Gelsinger in his place, who at one time was Andy Grove's protégé but left to eventually become the CEO of VMware, a leading cloud computing software company. Pat is smart, aggressive, and technical. He has already made sweeping plans to change the company or, in some ways, to relive its past. I wish him luck.

Other than attending Les's retirement party in 2003, I had no interaction with Intel Capital once I left in 1999. Les and I remain friends. We get to talk about the good old days while taking walks in Sonoma. Slowly, the people who had worked for me and were trained by me left

Intel. I am still in touch with most of them and have followed their largely successful careers.

— alumni can be a great asset

After I left Intel, I occasionally reached out to some of the senior executives and board members. Intel had been good to me, and I wanted to try to repay that. Paul Otellini and I were friends, and our homes were even close to each other in Sonoma. After Paul became CEO, I offered to help with strategy. He was not really interested. Only once did he ask me to review an Intel program. He did not like my conclusions and rejected them. I turned out to be right.

In 2013, when Krzanich became CEO and my friend Renée James became president, I sent them both a letter outlining a strategy for the company. In my opinion, the strategy could have made Intel much more of a player in the Internet market. Renée said that she and Brian read it. Perhaps Renée did, but I doubt that Krzanich did. There was no follow-up. It was disappointing. By that time, I felt I had made enough of an effort. I was done. Clearly, I have mixed emotions about not having enough of a positive impact on Intel. While I might not have had the answers, it was disappointing to me that I was not given an opportunity.

As for Intel Capital, Krzanich brought in Wendell Brooks, an Allen & Company investment banker, to handle mergers and acquisitions, which was previously part of Intel Capital. Brian was not happy with how Intel Capital had evolved. He pushed to downsize the international aspects and return to a much more strategic focus. Unfortunately, Brian's strategic insight was a bit off, as he was interested in things like drones, wearables, and augmented reality. These areas were too far from Intel's competencies. In 2016, with the retirement of Arvind Sodhani, Brooks also took over Intel Capital.

In 2018, Wendell also took over responsibility for the Emerging Growth and Incubation Group, another attempt to start new businesses within Intel. Intel continues to be very active in acquisitions, particularly in Israel. Wendell Brooks resigned from Intel in August

2020. Saf Yeboah-Amankwah, a twenty-five-year veteran of McKinsey, has become chief strategy officer, with responsibility for Intel Capital. Anthony Lin was promoted to head Intel Capital. For the first time since its formation in 1991, Intel Capital's head no longer reports to the CEO.

MY EXIT

When I left Intel in 1999, there would be no press release, so I called a few people to let them know. I had already agreed to join the boards of CMGI and World Online. On my second to last day, I called Mike Milken as a courtesy and told him I was leaving Intel. It was around 8:30 in the morning. Mike asked what I planned to do. I told him that I would advise companies and join boards. Mike said he wanted me to be his partner to make deals in health care. I told Mike that, while I really liked him, I was not interested in being his or anyone else's partner. He asked if I had lunch plans. I did not. He said he would fly up to San Jose (on his plane) and asked to meet me for lunch at the Fairmont hotel in San Jose at 12:30. I agreed.

While we had lunch, Marc Benioff came up to say hello. We both knew Marc. He told us that he had just left Oracle, where he was one of the top executives, to start a new company, called Salesforce.com. He briefly pitched the concept to us. Then he asked if we wanted to invest. Mike said he would put in few hundred thousand. I said that I would pass, because I did not invest in enterprise software. Clearly, this was one of the worst investment decisions of my life. Salesforce is now worth more than $200 billion, as much as Intel. I did agree to help Milken set up a company dealing with nutrition, called Miavita, and was on that company's board for a while.

On my last day at Intel, I had an exit interview with someone from the HR department in a small conference room. Typically, the HR representative attempts to understand why an employee has resigned. In my case, there were no questions about that. The main topics had to do with my returning Intel property, including my employee badge. Then, I was told that I had to leave my notebook computer behind. I had not thought about that. I had just assumed I would keep it. After all, it was not worth that much, and I had made billions of dollars for Intel. The HR person insisted: I had to turn over my computer. Had I made arrangements to buy another computer and transfer my personal files, I would have left it, but I had not. Keep in mind, I was at this point one of only thirty-five corporate officers, and I was being held captive until I returned my computer. I offered to compensate Intel for the computer but was told that was not possible. Finally, I said I would return the computer in a few days, after I bought another and transferred my personal files. Eventually, the HR person got approval for this. I had no intention of returning the computer, and I never did.

I then left the building. I had been working for Intel for almost fifteen years. I was both excited and sad. That afternoon, I drove to San Francisco, where I had rented a furnished apartment in the Cow Hollow district. I put my house in Palo Alto on the market. I was back home in the city of my birth. Remarkably, at the age of fifty-four, I had once again rejected a safety net. Although I was often frustrated being at Intel, the company did provide me with a platform I knew I could not replicate. But not knowing if I would survive prostate cancer incentivized me to move forward. I wanted to be on my own, no longer having to seek approval from my boss. Aesthetics mattered to me. I wanted to have a workspace that was beautiful. I actually hired a designer to help me create my office. I wanted a door, and I really wanted a view.

Fortunately, I had my Intel stock options. Intel stock had gone up more than a thousand times since I had joined in 1984. My options were worth a considerable amount, so I exercised them and held on to the stock for a year, selling within a dollar of Intel's all-time high.

Working at Intel had left me financially independent, but only because I had maintained the same lifestyle. I lived on my Intel salary. I saved and invested the money I received from my annual bonus and from exercising my stock options. Had I increased my spending as I increased my earnings, I would not have been financially independent. Worse yet, I would have had to find some way to replace a higher yearly earning number. This is an important lesson for all you young readers.

I OWE IT ALL TO DAN QUAYLE

I knew the valuations for tech stocks were crazy, but I thought it would continue for a while longer. My mantra was: "Give me one more year." One more year of growth in high tech. My objective was to double my net worth in that year. Happily, I did so and more. Most importantly, I sold everything I could in the first quarter of 2000, just before the tech stock crash. Timing, in this case, was everything.

It was former vice president Dan Quayle who helped me decide to sell my high-tech stocks.

Quayle had been George H. Bush's VP. Quayle was not considered the sharpest tool in the shed.

I was at a conference hosted by one of the leading investment bankers, Morgan Stanley, when I saw an old friend. He waved me over and said, "I want you to meet my friend, Dan Quayle!" Quayle and I shook hands. I asked him what he was doing these days, seven years out of office. "I'm a venture capitalist!" he replied. I knew at that moment we must be close to the end of the high-tech bull market. I started liquidating my high-tech stock holdings quickly.

Calculating the supply and demand, I was convinced that equilibrium would happen in early 2000. At that point, increases in stock price would have to be based on business performance. I knew there was also a chance that if supply overtook demand, prices would decline. That could lead more and more sellers to join the market, driving down the price still further, possibly even to collapse. Exactly that happened in March 2000. Had I known for sure, I would have taken action to benefit from the collapse in stock prices, buying options to sell (puts) or just selling stock short (selling stock you don't really own by borrowing the shares). But I was just happy to get out. I was not really a stock trader. All I wanted was to reach my own definition of financial independence. Once I obtained that, I was not willing to take much of a risk.

In 2003, *Forbes* listed me number eight on their Midas List of the top one hundred tech investors. When they called me for the interview (this would not be the first time I would appear on the Midas List), I told them I was no longer investing. This was one of their reasons for including me near the top.

CEO OF MY OWN LIFE

I was no longer at Intel. I had rejected the idea of becoming a partner in a venture or private equity firm. I certainly did not want a job at another corporation. I was not an entrepreneur and had no interest in starting a company.

What I really wanted to do was work with a few great CEOs that, unlike Andy Grove, might actually listen to me. I vowed that I would only work with people I really liked, who were doing things I cared about, in locations I wanted to visit, where I could have a significant impact, and where I would have the opportunity for some serious compensation.

To accomplish those objectives, I started my own company, which I called The Avram Miller Company. I described its purpose as business development and strategy. A friend, Larry Goldfarb, offered me office space at 50 California Street, right in the heart of San Francisco's Financial District. He even agreed to pay for an assistant. Larry and I did not actually have a business relationship. We had become friends many years before based on our mutual interest in playing jazz piano. Larry thought, correctly, as it happened, that if I had an office in his facility, many prominent people would meet with me there, and this would lead to introductions to Larry.

Joining the board of CMGI had been on the top of my list of engagements. I thought that Dave Wetherell was the most insightful CEO of any Internet company. In so many ways positive and negative, he was the opposite of Andy Grove. Dave was driven by intuition, insight, and passion. But unlike Andy, he was not process oriented and lacked many of the management skills needed to build out a major company. I think that that deficiency could have been backfilled. Had the Internet bubble not collapsed in 2000, or if CMGI had taken the action I recommended to raise a substantial amount of cash before the collapse, it is very possible that CMGI would be one of the leading Internet companies today. Dave and Eric Schmidt had discussions in the beginning of 2000 about merging CMGI and Novel, where Eric was CEO. Had that happened, Eric would have been CEO of the combined entity and Dave would have been the executive chair of the board. I recall speaking with Eric about this. Then came the crash of Internet stocks, and those discussions were on hold. Eric then became CEO of Google in 2001. Lucky man!

I joined the board of World Online. I had been very impressed with Nina Brink, the founder and CEO. I had lived in the Netherlands for five years, had been married to a Dutch woman for twenty-five years, and was still fluent in Dutch. I was very attracted to the idea of being involved with World Online.

A few days after leaving Intel, Richard Li, the CEO of PCCW in Hong Kong, asked me to join his board. I was reluctant. I did not understand what my responsibilities would be as a board member of a Hong Kong company, but Richard was very persistent. He said he would match the best deal I had been offered, which was CMGI. I finally agreed, but I also asked for a significant cash payment. I did not know, I said, how to evaluate the value of stock options in a Hong Kong company. And I'm certainly glad I did, because, as you will see, those stock options were worthless in the end—though, for a short time, they were exceptionally valuable on paper.

A good friend, Giorgio Ronchi, the former CEO of Memorex Telex, had started ETF, an open-ended venture fund in Lugano, Switzerland. They planned to eventually take ETF public on the Swiss stock exchange. I helped Giorgio structure ETF. Nina Brink joined the board. Ram Shriram, who was one of the first investors in Google (now Alphabet) and a current board member, also joined the ETF board. ETF did very well before the crash, increasing its value more than five times in one year. I tried to get Giorgio to sell some of the fund's assets to preserve sufficient cash in the event of a market correction, but he didn't agree.

The investment banking company, Lazard, invited me to become a senior advisor working closely with Richard Emerson, who later joined Microsoft as VP of corporate development. I had joined the board of King World the year before, at the request of Herbert Allen, and I was still on the board of the California Institute of the Arts. I resigned from the board of PluggedIn, the nonprofit in East Palo Alto where I had served as chairman for many years, and I joined the board of Equal Access, an international nonprofit based in San Francisco. I was one busy but happy guy.

Now that I was going to be working with one of the leading American, European, and Asian Internet companies, I began to formulate a concept. I would try to create an international venture capital consortium.

In addition to working with various companies, I was also making my own investments in early-stage companies. I made about twenty such investments in 1999. Most became worthless when the Internet bubble burst in early 2000. Thankfully, one of my investments, The Fantastic Company, went public before the market crash. I sold immediately because I was convinced that the market would crash. This more than made up for the other companies that failed. I hired an intern, Saif Mansour, to help manage my portfolio. I became his mentor and friend, a relationship that continues to this day. Later, Saif would start his own very successful company, Breakwater Management. I helped him get it going and then became an investor myself.

18. BURSTING BUBBLES

While I served on many boards and had several advisory positions, my main focuses were CMGI, World Online, and PCCW, and I want to share more about them.

CMGI

The topic of CMGI and its founder/CEO, Dave Wetherell, could be a book itself. David was one of the most creative and intuitive people I've ever met. He was a true visionary.

In 1986, CMGI was a sleepy little company close to Boston that sold mailing lists of professors to book publishers. Dave Wetherell, just thirty-two years old, became CEO after a leveraged buyout that year. In 1994, the company went public. Prior to CMGI, Dave had been a programmer. He had been studying the nascent World Wide Web, which Tim Berners-Lee had created in 1990. Dave was one of the first to realize the potential of the Internet and decided to start a company to sell books online. It was around the same time that Jeff Bezos started Amazon, but Dave was not aware of that effort. The company would be called Booklink.

Dave had managed to convince his reluctant board to allow him to develop this company. The budget, $90,000, was a lot of money for

CMGI at that time. Peter Mills was a friend of Dave's, and they had explored creating a company before, so Dave turned to him for help with Booklink.

Dave realized that for Booklink to be successful, it would need some easy way for consumers to navigate, so he had Booklink first develop a browser. Netscape had yet to launch Netscape Navigator. He had them develop a browser first. Then as luck would have it, Dave found himself seating next to Steve Case, the CEO of AOL, on a flight. Case was looking for a browser. One thing led to another, and CMGI sold Booklink to AOL for $34 million in AOL stock, but I believe they cleared $72 million by the time the stock was actually sold. This money was then used to create a venture capital fund called @Ventures. Initially, Peter Mills and Dave Wetherell were the general partners and CMGI was the only limited partner. Later, other general partners were added.

One of their first investments was in a start-up called GeoCities, which was one of the very first social networks, founded by David Bohnett in November 1994. When Yahoo bought GeoCities for $3.56 billion in January 1999, it was the third-most-visited site on the web. GeoCities had gone public the year before at $17 a share and traded as high as $100 before Yahoo's acquisition.

Intel was considering an investment in the first round of GeoCities, which CMGI's @Ventures was leading. A member of my team, P.S. Kohli, organized a meeting for me with Peter Mills and Dave Wetherell. Fortunately, we did decide to invest in GeoCities. During the meeting, Dave told me about CMGI, and I was very intrigued by the concept. I was also extremely impressed with Dave and his understanding of the Internet.

Dave's concept for CMGI was rather novel. Besides @Ventures, CMGI created new Internet companies directly in which it would have controlling interests but in which there would also be other shareholders, including employees. These companies could eventually go public. This way, CMGI could offer founder shares to the key employees of these start-ups.

I was so impressed after meeting Dave that I proposed that Intel invest directly in CMGI. This was very unusual for us, since CMGI was a well-established public company and not a start-up. I saw an investment in CMGI as a way for Intel to get access to Dave's strategic mind and insight while creating many additional investment opportunities. In December 1997, Intel bought 5 percent of CMGI for $11 million. At that time, Intel did not want its employees to serve on the boards of its investments. I was given observer rights and began to attend all the CMGI board meetings. Intel eventually sold its position in CMGI for $388 million, although I suspect Intel's position might have been worth a few billion dollars before the crash of the Internet stocks.

Dave and I developed a close friendship. We are no longer in touch, and I miss our friendship. I would stay at Dave's house when I visited CMGI. I became close with his then wife, Celeste, and knew his children well. Dave, like me, played the piano. We bonded over that. He had recently bought a Fazioli piano, which was and is the world's best piano. I bought one myself in 2001.

I suggested that Dave strengthen his board with someone from the communications industry, and proposed my friend Bill Berkman, whose family had made a significant fortune in telecommunications and who had also been the second-largest shareholder of TCI when it was sold to AT&T. Bill was just thirty-three at the time. Besides his solid understanding of communications technology, Bill was extremely smart about financial engineering. Dave brought Bill onto the CMGI board.

After I left Intel, I really wanted to join the CMGI board and especially work with Dave. A few weeks before I resigned, I told Dave about my plans, and he did ask me to join his board. Besides offering me a substantial amount of stock options, he said that I could, if I liked, also join a few of the boards of CMGI's subsidiary companies. I would get additional shares in those companies. While the stock was important to me, I really just wanted the opportunity to work with Dave.

I joined CMGI's board in April 1999. The company's market value was about $10 billion. One year later, it had risen to $41 billion, worth almost as much as General Motors. My options vested monthly, fortunately. At its height, CMGI had seventy investments, twenty subsidiaries, and five thousand employees.

Not counting @Ventures, CMGI owned a controlling position in Lycos, one of the major portals, and many advertising serving companies, like Engage. They had one of the first server hosting companies,

NaviSoft. Later, CMGI would acquire AltaVista, the leading search company from Compaq.

Dave's model for CMGI worked like this. Dave would either find a start-up or create one, with funding from CMGI. The start-up's management team would take founder shares in the start-up so they were cheap. CMGI would fund the start-up and might bring in other investors, especially if they were strategic. CMGI would control the board, and many CMGI board members would also sit on the start-ups' boards. I myself was on several. Though this could create conflicts of interest, there were only a few lawsuits. As these start-ups increased in value, the value of CMGI would also rise. When CMGI needed cash, it could sell shares in the subsidiaries that had gone public. The constantly increasing value of Internet stocks meant this was like owning a printing press for money—except we eventually ran out of ink.

— the right thing at the wrong time is the wrong thing

One of the CMGI start-ups, ICAST, was like YouTube, but six years too early. Timing is undoubtedly one of the determining factors in success. I like to joke that start-ups are either too early or too late. Being too early works out if you can wait out the trends. Being too late seldom works. Unfortunately, the Internet meltdown made that waiting impossible.

Dave understood that there would be user-created video content and wanted to create a site for that. He also thought that the same site could stream professional content, envisioning a combination of what is now YouTube and Netflix. Dave hired Neil Braun, then president of NBC Television, to head the new company. CMGI committed to funding ICAST with an initial $100 million. I joined the ICAST board. Soon after its launch, tensions brewed between Neil and Dave. The CMGI board and Dave had growing concerns about ICAST's spending. I don't know why I didn't realize ICAST was too early. Eventually, Neil was let go. Neil sued CMGI, seeking $50 million. It got really messy, I remember, but Neil's legal actions did not succeed. Margaret Heffernan, CEO of one of the smaller CMGI companies, ZineZone, became ICAST's

COO and acting CEO. By the end of 2000, ICAST would be no more, falling victim to the Internet crash like so many others.

ICAST, as it happened, played a significant role in my personal life. On March 9, 2000, at his home in the Brentwood neighborhood of Los Angeles, I was having dinner with Mike Ovitz, the former founder of Creative Artists Agency and then president of The Walt Disney Company. Mike reached out to me sometimes to get my perspective on what was happening in technology and media. We ate sushi, which Mike loved to the extent that he owned a sushi restaurant. That day, I had given a presentation at the Milken Economic Forum. I promised Margaret that I would show up for ICAST's launch party at the Skybar at the Mondrian Hotel in West Hollywood. The day before, I had met with the former CEO of American Cybercast, Sheri Herman. She had a new venture. I suggested she come to the ICAST launch party so I could introduce her to a few people. I was supposed to show up at nine p.m. and I almost didn't make it, lost in conversation with Mike, but I felt obliged. Sheri, her fiancé, and one of her friends showed up at the same time.

Sheri introduced me to her friend Deborah Neasi. We shook hands, talked for twenty minutes. Within six months, we would be living together and then marry in 2003.

BEFORE THERE WAS GOOGLE, THERE WAS ALTAVISTA

The search engine AltaVista was developed by Digital Equipment Corporation at their West Coast R&D center and launched in 1995. Compaq had acquired Digital, including AltaVista, in 1998. On June 29, 1999, CMGI acquired 83 percent of AltaVista from Compaq for $2.3 billion, of which $2 billion was CMGI stock and the rest was a three-year note. As part of this transaction, Compaq got a CMGI board position, filled by Bud Enright, vice president of business development at Compaq. Bud became a dear friend. He probably left the CMGI board when HP acquired Compaq in September 2001.

The management of AltaVista wanted it to become a portal like Yahoo, Lycos, Excite, and AOL. Google wisely decided to focus on search even before anyone had figured out how to make money that way. Search would come to undermine the importance of having curated sites like portals. The AltaVista board approved the change in focus to becoming a portal. I don't remember my position on this.

AltaVista was also working on an IPO, targeting April 2000. This timing was horrible. The market for Internet stocks tanked in March, and we withdrew AltaVista's IPO.

Google was started in 1998. I recall a discussion during an AltaVista board meeting sometime in 1999 about acquiring Google. As I recall, Kleiner Perkins had indicated we might be able to buy the company for something like $80 million. The AltaVista board felt that was too much. Of course, by 2003, Google was looking at purchasing AltaVista for just 2 percent of Google's stock. Instead, the AltaVista board decided to sell to Overture for $140 million. Yahoo bought Overture soon after, and AltaVista's search was phased out and replaced. I probably left the AltaVista board in late 2000 or early 2001. AltaVista had many important patents. Google acquired the rights to those patents right before their IPO. Google paid Yahoo 5 percent of the company for the right to use those patents. This probably resulted in more than $10 billion in profit.

In early 2000, CMGI stock had gone through the roof, like many Internet companies; we were now worth something like $40 billion. My friend and fellow board member Bill Berkman and I were concerned about the high valuations. I was starting to sell all my high-tech stocks.

— *cash is a strategic asset*

Bill and I proposed a secondary offering to the CMGI board to raise a few billion dollars. Dave wouldn't do it; he felt that CMGI stock was too valuable to sell. Since he was convinced it would increase in value, the discussion was pretty contentious. Cash was a strategic asset, I argued. If there was a correction in the stock market, we could wait it out and even buy other companies for cash, at a discount. In the end, we compromised. CMGI owned about a billion dollars of Yahoo stock, received from the sale of GeoCities. We agreed to sell that stake and raised the cash. Had we not done so, CMGI might have not survived the crash that followed. In discussing selling CMGI stock with Dave, I asked him if he thought I would be a fool to sell my stock. He said yes. "Then I am a fool," I said, and I sold all of my vested shares. I was not alone. Other board members sold their CMGI stakes. Had Dave agreed with the secondary offering, I would not have been able to sell my shares. On the other hand, had CMGI raised that cash, it could have picked up a lot of companies for almost nothing after the bubble burst.

The bursting of the Internet bubble was like the low tide, revealing all the rocks and garbage of the Internet. AltaVista pulled its IPO. Many public CMGI subsidiaries saw their prices decline substantially, and CMGI, just the sum of its parts, declined as a consequence. Eventually, CMGI traded at 2 percent of its highest value. Dave, who had been a multibillionaire, was down to his last $100 million. What a ride!

CMGI's board was displeased with Dave for buying the naming rights to the New England Patriots stadium. CMGI had agreed to pay $7.6 million per year for fifteen years; in exchange, the stadium was named CMGI Field. CMGI sold no products to consumers—or really to anyone. It was just a collection of wholly or partially owned companies that marketed under their own brands. Frankly, I always thought that a great early-warning indicator of failure is when companies buy naming rights or build new headquarters. CMGI as a company had no direct connection to consumers. So why would it make sense for CMGI to acquire the naming rights to a stadium? Because it did actually have a consumer product. It was its stock.

For this and other reasons, the board had growing concerns about Dave's ability to manage the company during such challenging times. But none of the board members, including me, was willing to take action. Looking back, I'm not sure that replacing Dave at that time would have been the right thing after all. At the time, though, I felt it would be. I believe that a board's primary responsibility is to make sure that the company has the right CEO. We were failing at that task because none of us wanted to handle the consequences. That's when I decided to leave the CMGI board, as did several of the other board members. By 2002, there was a new board, new CEO, and the company was restructured as an operating company. Dave stayed for a while as chairman and then left. He subsequently moved into health care venture investing, founding Biomark Capital, which I believe has been successful. @Ventures continued under Peter Mills and raised several funds. I still regret that CMGI had not raised more capital during the boom years. We would have been in an excellent position to pick up the survivors.

Some have speculated that CMGI could have been Google if not for the Internet stock crash. I tend to agree. Dave had identified all the key areas from Booklink (commerce and browser), GeoCities (social media) Lycos (portal), Navisoft (hosting), AltaVista (search), and ICAST (video sharing).

PCCW

Who would have thought that in 1963, a skinny, eighteen-year-old merchant seaman who fell into Hong Kong Bay would later end up on the board of directors of one of the most important companies in Hong Kong?

Just before I left Intel, I shared my plans to leave with a select number of CEOs of companies with which I worked closely, including Richard Li. I mentioned to Richard that I would be joining the CMGI board. Richard asked to meet for lunch in San Francisco the following week, as he would be in town and had something he wanted to discuss with me.

When we met for lunch, Richard told me right away that he was in the process of acquiring a small company that was already listed on the Hong Kong Stock Exchange. He would then merge his PCG into that shell so it would become a public entity. He wanted me to join the board of the resulting company, which would be called Pacific Century CyberWorks (PCCW).

I was not so keen. I wasn't personally close to Richard. In Intel's work with PCG, I had worked mostly with Michael Johnson and George Chan. While Richard was very intense, with tremendous energy, smart, and driven, he wasn't much of a listener. I knew little about how corporations functioned in Hong Kong, and I realized that even if PCCW were a listed company, Richard would own the majority of the shares and totally control it. So, I told Richard I felt already over-committed with my new engagements.

RICHARD LI MAKES ME AN OFFER I CANNOT REFUSE

Richard was really insistent. He said he needed my strategic mind, but I suspected he mostly wanted to gain recognition and legitimacy for his new venture. Richard decided to make it about the money. Richard asked me: What was the best deal I had negotiated for myself? I told him that I had been granted options for 0.5 percent of CMGI. Richard said he would match, offering 0.5 percent of the outstanding shares of PCCW. As it turned out, and for a number of reasons, the offers were not really equivalent. When I ultimately saw the actual offer, it had a vesting schedule that started at 10 percent of the total offered, with the rest vesting over a nine-year period. I thanked Richard, but I told him that I really did not understand how the Hong Kong stock market worked. So, in addition, I asked him to pay me a significant sign-on fee. He agreed and actually wired the money the next day, before we had agreed to all the details. The agreement was written as compensation for my advising the company. It provided that I would join the board and remain there for the term of the agreement if requested by the company. I am not sure why we did it that way.

THE HIGHEST-PAID ADVISOR IN THE WORLD

After the agreement was signed, I wanted to join the board right away, which I thought would give me the best chance to help Richard and PCCW. I was tasked to help PCCW develop its venture group, which I did. I actually pushed Richard to make the board appointment. I think he wanted to delay, because he knew that once I joined the board, our agreement would have to be made public. It was actually worse than that. The Hong Kong Stock Exchange required that an independent organization review our agreement, including compensation, and determine whether it was fair. Goldman Sachs was hired to lead this effort, and the compensation study took about four months.

In the meantime, PCCW's stock was going through the roof, like most Internet stocks. Eventually, PCCW was one of the most valuable companies in the world. I remember an event attended by Richard and Masa Son, the incredible Japanese entrepreneur, at which they argued with each other about who had the most valuable company. By the time the board received the fairness report on my compensation, my stock options, assuming they were all vested, were worth $90 million: 63,201,097 shares at a strike price of 2.36 HK dollars. The *Wall Street Journal* picked up the story, and I became notorious: the highest-paid advisor in the world. While the compensation was a factor, I was even more captivated with the idea of developing an international consortium of key Internet companies in Asia (PCCW), Europe (World Online), and the United States (CMGI) and put together an international corporate venture investment structure. I want to create something on the scale of Intel Capital.

My compensation package had to be approved at a special meeting of PCCW shareholders. That's when I heard from a friend at Intel that something really bad was about to happen. Two key Intel executives who had been tasked with managing Intel's PCCW shareholder role had decided to cast Intel's votes against me. Intel owned about 15 percent of the total shares of PCCW. I couldn't believe it. I was so angry. How would this look to the outside world? It would be bad for everyone. I couldn't see how it would help Intel, because they didn't have enough votes to stop the approval. Intel's disapproval would reflect badly on PCCW. And it would be very bad for me to have my former employer vote against me.

I don't know if the two executives had decided to vote against me out of some kind of vindictiveness, given the high value of the compensation, or if they felt there would be a conflict of interest. I called Les Vadasz. He had not known about this and was livid. The Intel executives had never discussed this with him. He said he would not have supported it, as they must have known. I suggested to Les that Intel could just decide not to vote, explaining they felt it was a conflict of interest to vote on this matter since I had recently been employed by Intel. That is what happened. For many years, I avoided the two Intel executives who had planned to vote against me. Eventually, I got over it, and today I'm in contact with both.

— don't count the money until you cash the check

I joined the PCCW board on January 14, 2000. Although my shares were worth $160 million in theory, in reality, while I could have only sold the 10 percent that had vested when I signed the agreement with PCCW, I did not feel it was right to sell those shares so soon after joining the board. I wanted to sit on the board for at least a year before I sold any stock, which would have been in August 2000. By that time, the Internet bubble had burst, and my PCCW options were underwater. I never made a cent from the stock. My initial concerns about the Hong Kong Stock Exchange were proven right. So much for being the highest-paid advisor in the world.

BUYING THE HONG KONG TELEPHONE COMPANY

Cable and Wireless started off as a British telephone company, with roots back to the nineteenth century. They built communications companies in British colonies, including Australia and Hong Kong.

When PCCW was established, Cable and Wireless HK was the leading telephone company in Hong Kong. Cable and Wireless was downsizing, selling off parts of its international holdings. So, less than six months after PCCW's founding, the opportunity presented itself to buy what was, in effect, the Hong Kong phone company.

— buy tangible assets with Internet dollars

PCCW's market value was approaching $50 billion, and it had almost no revenue. I personally thought that the valuations of Internet companies were crazy, so the chance to change some "funny money" of Internet stock for real assets was very attractive. PCCW could also use the company's infrastructure, of course, to offer Internet services to Hong Kong. PCCW bought the majority of Cable and Wireless' holdings in Hong Kong for about $38 billion in cash and stock. They now had more than three million fixed line, one million mobile, and five hundred thousand Internet customers, plus a fiber-optic network covering most of Hong Kong. PCCW was just ten months old.

After the Internet market crashed in 2000, most of the value of PCCW was its ownership of the telephone company. It had taken on a lot of debt as part of its acquisition and suffered under pricing pressure in the telephone business, but, unlike so many other Internet companies, PCCW was alive. Richard gave up his position as CEO but continued as board chair.

PCCW's venture arm, CyberWorks Ventures, which I helped set up, made eight venture investments in the first quarter of 2000. Given the market crash soon after, I would guess they turned out worthless. But PCCW had sold its very expensive stock to fund such investments.

Michael Johnson decided to work on branding an Internet service and complementary television network, offering original content, called Network of the World. He bought the domain Now.com from *NOW Magazine* located in Toronto for many millions of dollars. Michael moved to London, where he set up a studio just outside the city. Michael, who at one time had made movies, was in his element. PCCW was now positioned as a worldwide Internet company. Together with the promise of penetrating the wide-open Chinese market, that put the company in the super leagues and helped drive its sky-high valuation. But by December 2000, the party was over for NOW and most Internet stocks, and Michael was removed from his position. NOW's budget was cut from $300 million a year to $3 million. PCCW's business was primarily the HKT acquisition.

In 1999, Intel merged its ownership in Pacific Convergence Corporation with PCCW. At one time, it owned about 13 percent of PCCW. Intel also directly invested in PCCW and became a pretty large shareholder. Intel's total investment in PCCW of a bit more than

$60 million and would eventually return more than $400 million when it finally sold its share. Its position was worth more than $3 billion at the peak.

Seated: Richard Li, Dave Wetherell, and me

BRINGING CMGI AND PCCW TOGETHER

Rather than figuring out how to bring CMGI's potentially important Internet assets to Asia, I proposed to Dave Wetherell that CMGI explore working with PCCW. He was interested. I discussed the idea with Richard Li, and he, too, was interested. I arranged a meeting with David, Richard, and key people from both companies. It started with a meeting at PCCW in Hong Kong, and continued on Richard's yacht, which we boarded in Phuket. We spent a few days discussing ideas and drinking a lot of champagne.

To solidify the relationship between the two companies, they exchanged newly created shares worth $350 million. Soon after, CMGI acquired an additional $500 million PCCW shares from Cable and Wireless, once again by issuing new CMGI shares. In a sense, issuing shares is akin to printing money.

A joint venture between the two companies was established in Hong Kong. CMGI put in all of their Asian assets, which involved eighteen different entities. The joint venture would be dissolved about six months later, after the Internet market collapsed. I was the advocate of the cooperation between PCCW and CMGI. Had the Internet stocks not crashed in 2000, I think this would have worked out very well for both companies.

I remember attending the CMGI annual board meeting in 2001, after the crash. Some shareholders attacked us—and especially me—for making the $350 million investment in PCCW. He did not really understand that, in fact, we had a gain, even though we showed a paper loss. We had issued new CMGI shares in exchange for the PCCW stock. While PCCW had declined in value, CMGI went down even further; PCCW lost more on its CMGI stock than we had lost on our PCCW stock. Of course, deciding whether or not CMGI made a good investment would have had to account for what CMGI would have done instead (that is, the opportunity cost). It is all very complicated, certainly not something I could have explained at a shareholders' meeting. I sat quietly while the shareholder yelled at me for having done something that had benefited him.

In 2002, I decided to leave the PCCW board. I was leaving all the boards on which I had served. None of the companies were particularly in trouble, but neither were any particularly interesting. The crash in 2000 had shattered my dream of putting together an international group of leading Internet companies around the world. I no longer had much to offer and no longer had much interest. Traveling to Hong Kong every three months for board meetings was a lot of effort, and Richard seemed to have little interest in spending time with me.

About a quarter of my options were vested by then, but they were still underwater, meaning that the strike value was higher than the current market value. I could have kept them, but I offered them back to the company. Since then, I have had little contact with anyone at

PCCW. My only contact with Richard is a Christmas card I receive by email every year. I did hear from Michael Johnson occasionally before his untimely death in December 2020. I am in contact with a number of the people who were part of the PCCW Venture Group I helped to form. Evidently, things worked out there, with some significant investments in major Chinese Internet companies.

WORLD ONLINE

I learned about World Online (WOL) in July 1998, at the Allen & Company Sun Valley Conference, in conversation with my old friend Mickey Schulhof, the former president of Sony of America. He was chairman of the board. WOL was an Internet service provider based in the Netherlands. Mickey told me about the CEO, Nina Brink, a Dutch Jew and serial entrepreneur. I asked him to arrange an introduction.

I wanted to meet her for three reasons. First, I thought World Online might be an attractive potential investment for Intel (it was). Second, my son Asher and his wife were planning to move to Amsterdam so that Asher could do research for a book he was writing about the Holocaust, and his wife was looking for work there. Third and finally, I was looking for board opportunities after I left Intel, which I intended to do the following year. Sitting on the board of a Dutch company was very intriguing to me as I had lived in Holland, had many friends there, and still spoke fluent Dutch.

Nina and I met in San Francisco, and we hit it off. Her intelligence and how she described the WOL opportunity impressed me, as did the fact that Mickey served on her board. Mickey would not be on the board that much longer, as it turned out.

I introduced the idea of an investment by Intel in WOL to George Coelho, who was responsible for Intel's international venture investments. World Online had close to two million users and was growing rapidly. In Europe, its presence was as significant as AOL's penetration of that market. Though the company primarily dominated the Dutch market, without a substantial presence in other European countries, it had the potential to expand in those countries as the Internet began to take off. I told George that I did not want to be personally involved in any of the discussions or negotiations around a potential investment in WOL. Intel invested $67 million in June 1999, by which time I no longer worked at Intel. For that money, they owned 10 percent of WOL, valuing the company close to $700 million. My daughter-in-law did end up with a job there, but I was not involved in those discussions other than making an introduction.

When I told Nina of my plans to leave Intel, she asked me to join her board. I did so in April 1999 and was pretty active with the company in terms of giving advice. Most of the company stock was controlled by members of the board: one from a wealthy Dutch family and another who managed the money of the Sandoz Family Foundation.

I introduced Nina to Dave Wetherell of CMGI, and discussions about collaboration began, but the Sandoz board member was against the idea. Nina also met my friend Larry Goldfarb of LRG Capital, with whom I shared office space. Larry and Nina explored setting up a venture group for WOL.

Nina had most of her net worth tied up in WOL, which made her concerned. Evidently, Larry's firm, LRG Capital, bought some of her shares for a combination of cash and options on those shares. That way, Nina retained upside on about half of her shares. I didn't know about this, and I don't think any other board member knew about it, either. The transaction was not something we were required to approve, but it turned into a big problem once WOL went public.

THE BIGGEST IPO IN THE HISTORY OF THE AMSTERDAM EXCHANGE BLOWS UP

The value of Internet stocks was only going up in 1999. World Online decided to go public on the Amsterdam Exchange. Goldman Sachs and ABN AMRO Bank were the investment bankers. The company went public on April 2, 2000, just as the Internet bubble was breaking. The company sold $2.9 billion in stock at a total valuation in excess of $12 billion. Usually, in the United States, companies do not sell many shares at their IPO, helping to create demand and pushing up the price. But the Amsterdam Exchange required a minimum offering of 20 percent of the company's shares. In retrospect, this was helpful; even though the value of WOL declined after the Internet stock crash, all that cash WOL had raised put a floor on the value of the shares and made the company an attractive acquisition target.

The board met with the bankers in a meeting I will never forget. We were twenty-two times oversubscribed; the bankers told us. That meant there were potential buyers for twenty-two times the amount of stock we intended to offer. All IPOs set an indicated range of price until just before the offering, when a company's board and bankers will meet to decide on the exact price within that range. Given the strong demand, we decided to price at the top of our range, $42. The stock was already trading on the UK gray market at $72.

The stock opened at $50 per share. We all expected it to rise from there, but the Amsterdam Exchange couldn't handle the volume. Their computers kept crashing. Then there were rumors that people with significant positions were selling. Many in the room thought it might be Larry Goldfarb. Others thought it was Mickey Schulhof. Neither Larry nor Mickey had signed lock-up agreements like the rest of us. (We couldn't sell our shares for six months.) On the first day of trading, 57 million shares changed hands of the 64 million that were sold at the IPO. Larry denied that he had sold.

That night, we had a big party to celebrate. But many of us were worried that the stock price would not be sustained.

World Online stock continued to decline, mainly for two reasons. First, the stock had been hyped up. Many buyers, including many retail buyers, had just intended to flip the stock once it had gone up a lot. When it did not, they sold. Second, the market for Internet stocks in general was crashing.

Making matters worse was the fact that Nina had sold stock at $6 per share earlier in the year. This was not adequately disclosed in the prospectus, it turned out. Instead of saying that Nina had sold stock to LRG Capital, it said that the stock had been transferred to LRG Capital. I had read the prospectus many times, and carefully, since, like the other board members, I had to sign off on it. But I did not notice this.

Now there were lawsuits and investigations. World Online's offices were even raided. The board took action, removing Nina from management. It was a challenging time for her. The press ran a lot of very negative, sometimes even personal, attacks on her. We continued to grow the business. Like almost all Internet companies, though, we were navigating difficult times.

About a year after the IPO, we sold the company to Tiscali, the most significant Italian Internet service provider. They needed cash to grow, and we had it. The new CEO of World Online, Jim Kinsella, became CEO of the merged company, while Tiscali's founder became the chairman of the board. I would not join the board of the merged company.

Strangely, the night that I could first sell my shares in Tiscali was the night before the CMGI Annual Meeting. I was up that whole night on the phone with people at Goldman Sachs in Milan, selling all the shares I had received from the sale to Tiscali. I was totally beat the next day, even regrettably dozing off during one part of the meeting. Dave Wetherell was, rightfully, very angry with me for that.

Nina and I stayed friends. She has some investments in Israel, so I see her often in Tel Aviv.

THE FORMER AVRAM MILLER

In 2002, just two years after leaving Intel, I decided to give up my office. I was no longer on the boards of CMGI, PCCW, or World Online. My advisory agreement with Lazard had ended. Most of my early-stage investments had failed. Nevertheless, in my two years after leaving Intel, I had doubled my net worth. While I was not "private-jet rich," I was comfortable. I continued to work with a few early-stage companies such as Heavy.com, but my heart was no longer in it. I didn't want to be a dilettante or, worse yet, work my way down. Many of my friends in similar circumstances started their own venture funds, but I had no interest in that.

— the half-life of a technology visionary is six months

I knew that my status would be significantly downgraded by leaving Intel. Now, the companies on whose boards I had worked, while still functioning, had little influence in the computer industry. I would sometimes introduce myself as "the former Avram Miller." But I did not mind as my ambitions were satisfied. My bout with cancer had altered my perspective about life. I was no longer willing to spend all my time working.

My children were grown, married, and on their way to creating their own families. I wanted to continue my piano studies. I had a teacher three hours a week. I composed music and thought that I might want to compose the soundtrack of a movie. Travel became and still is a big part of my life. Deborah and I travel the world and spend at least four months a year in this pursuit. Staying healthy and fit became a priority.

I started working on a book about residential broadband. I soon discovered that I lacked the skills or discipline to write a book. I had a friend who was an investor in the company that owned WordPress. He introduced me to its founder, Matt Mullenweg. We had lunch. Matt was so young, maybe twenty years old at that time. I was used to having meetings with people young enough to be my children, but now I was having a meeting with someone young enough to be my grandchild. I asked him to give me a lesson on writing a blog, which he did, and I started my blog, *Two Thirds Done*. I was sixty years old. I figured I could likely make it to ninety. The blog is still called *Two Thirds Done*, even though I'm now halfway from sixty to ninety. Sometimes I think I should change the name to something like *Almost Done*. I learned a great deal from writing the blog that helped me to write this book.

My friend David Agus, a renowned cancer scientist, kept encouraging me to be more active. He felt that I had still a lot to offer. Eventually, I did get seriously engaged in the development of a breakthrough in the treatment of obstructive sleep apnea. I helped create a company called Sommetrics and was, at one time, its largest investor. The company's device is now undergoing clinical trials in order to get FDA improvement. Assuming it is approved, it should have a major impact on the lives of so many.

But I was tired of working in business. I decided I would focus on my own creative pursuits and go back to learning. I delved deeply into physics, especially quantum mechanics. Deborah actually arranged for me to have a private session with the physicist and author Sean Carroll from Caltech. I studied several hours a day for three months in preparation for my two-hour session with Sean, which actually turned into three hours. While the meeting was terrific, the four months of study were even better. I reached a level of comprehension that I found very stimulating. I also became fascinated with genetics and took several online courses. I started writing music again, and I learned how to use tools like Logic to write and arrange music. I continued to take jazz piano lessons. But I found and continue to find it challenging to work alone.

— advising early-stage companies is an exercise in frustration

I continued to advise early-stage companies, primarily to help friends and keep current with technology. Some of these companies were in entertainment, while others developed communications technology, and still others were in medical device technology.

While I enjoyed working with these companies, I saw no real impact from my efforts. I decided that I was done. I was retired. My wife said, "No, you are not retired, you're rewired." But she also noted that even after I retired, I continued to get involved with early-stage companies. She was right; during that period, I advised more than twenty early-stage companies. Among them, two have had successful exits, and another six have had substantial unrealized gains. The rest are either not doing so well or are dead.

Advising early-stage companies is like angel investing, except you use your time instead of your money. Most early-stage companies fail, so you have to be involved with enough of them to have a few winners. In the beginning of 2020, I stopped any advisory activities to focus on completing this book.

As I met with CEOs of early-stage companies, I realized that I was mostly repeating the same concepts and lessons. That helped inspire me to write this book. Not only could I document my journey and views on the evolution of the computer industry, but I could also share lessons I've learned. Lately, when people have asked me for advice, I could say: "I wish I had time to help you but am so busy writing my book." Perhaps now I can just say, "Here is a copy of my book. I think it will be helpful to you."

EPILOGUE

REFLECTIONS

Looking back over seventy-six years of life is daunting. So much of the world has changed, and yet so much remains the same. Every life is a melody that is played only once but creates harmony and counterpoint with other lives, other melodies. We start out young, and if we are lucky, we get to be old.

Writing this book has taught me much, some of which was not easy to accept at first. When I started writing, it was with the hope that I could share my experiences and understanding with others so that they might benefit from my journey. I did not realize that it would also change how I viewed my own life.

My career was driven by curiosity, creativity, and the desire to have an impact. Financial benefit, power, and notoriety, while they played a role, they were never key drivers. For me, the purpose of life consists of two elements. The first is to appreciate existence. It does not matter if one is religious or not; we can all be inspired and be in awe of the beauty of the universe and all that it provides. The second is to leave a positive mark, to know that you added value, that somehow the world is a better place because you were in it.

This book deals mostly with the second element. I had the fortune of being able to make some meaningful contributions, both during the time I worked in medical science and then in the years I spent in the computer industry. Although much of what I tried to do was not successful, I think nothing was a failure per se because I learned so much in the process. While I have written much about the importance of luck, intuition, creativity, and humor, I think there is another ingredient that is equally as important—tenacity. It is the sibling of courage because courage will often lead to failure.

Luck played a major role in my life, as I suspect it does for all of us. There is a sort of randomness that most of us experience. But we shape our responses to what comes. We start out with a deck of cards that was dealt to us by our parents and their circumstances. How we play the hand is largely our choice. Our values shape our decisions. Generosity, compassion, and integrity can conflict with ambition and desire.

As we travel through our careers, we are presented with opportunities to learn and become more capable for our next task. Learning is critical to success. Often, it is our failures that provide our greatest lessons. I certainly had my share of them. However, I am also fortunate to have had significant successes.

I have always lived in the future. The future was a comfortable place for me, and I loved being involved in creating the technology that would power it. At this point in my life, my contribution to the future of technology is to help document its past. I have more knowledge and observations dealing with what transpired during the time when I was active that I want to share, but it is time to bring this book to an end. Therefore, it is my intention to write a series of essays and lessons that I will publish online. Some may be there already as this book is published, at WildDuckFlight.com.

This book does not really cover my personal life. But my family played a big role in my decision-making. It was important to me to provide my family with security, which limited some of my choices. For instance, starting my own company. However, my life choices also put my family under a great deal of stress, particularly with respect to moving not only from city to city but from country to country. Like so many, I was torn between wanting to be a great father and husband and the many challenges of my career.

By 1993, I was on my own. Arianne and I decided to end our marriage but not our friendship or family. My children were all living away from home, attending school. For the first time in twenty-five years, there was no one who expected me to be home. I spent all my time working. This was the period when I focused on the creation of residential broadband. I was totally driven by a vision for a world in which everyone and everything would be connected, and I could devote all my time making this happen.

My children grew into wonderful human beings. My two sons, Adin and Asher, have devoted their careers to the nonprofit sector and serve as executive directors of nonprofit organizations. They have wonderful families, and each has two sons. My daughter, Dafna, who left us way too early in 2019, was an A-list Hollywood event planner.

In 2000, I met Deborah. She is my constant companion. Of her many gifts to me was teaching me to live in the moment. I am still learning that. We have traveled the world. We decided to live primarily in Tel Aviv, although we still have a home in Sonoma. In 2009, we meet a young girl named Karma in Bhutan. It was her twelfth birthday, and we "adopted" her. Eventually, she studied graphics arts in Toronto, where she now lives and works. She, too, is a wonderful and talented human being.

My book is over. Thanks for reading it. I hope it encourages you to be a wild duck. Raise your wings and take flight. Leave the other ducks behind and soar.

ACKNOWLEDGMENTS

Many people helped me while I was writing the book. First and foremost, I must thank Deborah Neasi-Miller, my amazing wife. Her encouragement was fundamental in my ability to keep at the task of writing this book. After twenty-one years, Deborah knew and could remind me of many of the stories that I incorporated in the book. Her patience with my constant discussion of the book demonstrates that her love and generosity know no bounds, including doing the graphics for various charts in the book.

While researching the book, I conducted more than seventy interviews. I want to single out a few people. At the top of my list must be my former boss, mentor, and friend, Les Vadasz. Les, as always, was there for me with both encouragement and criticism. We had numerous discussions in person and over Zoom as well as extensive email exchanges. Les never held back when he had criticism.

Mike Richmond was the person who originally introduced me to Intel in 1984. He helped clarify several important Intel decisions. Mike's suggestion that I work with his son David Richmond, a copywriter and researcher, really paid off for me. David helped me organize my research work and provided important preliminary editorial advice. I may not have written this book without this early help.

Claude Leglise clarified important details about the early development of the Intel microprocessors business. Steven McGeady helped me understand key developments within the Intel Architecture Labs and, in particular, Intel's efforts in video compression. John Calhoun clarified many decisions taken by the Intel System Group. Sudhir Bhagwan helped me understand the early days of the Gemini program. John

Miner gave me insight into the Barrett CEO period. Mark Christensen clarified Intel's efforts in communication and networking.

Thanks to Matthew Cowan, Sriram Viswanathan, Bart Schachter, Davin McAndrews, Jacob Tanz, and Ravi Jacobs for the help on the Intel Capital section.

Steve Maine was constantly there for me as I wrote the section of the book dealing with broadband communications. Steve Craddock and Teri Lasley helped me remember the details of the early cable modem trials. Mari Vecchio helped me increase my understanding of both Time Warner Cable and later AOL's thinking in the early development of broadband.

Several people read early versions of the book and provided very valuable feedback. Chief among these were Karen Alter, Hilton Barbour, Matthew Cowan, Dane Elliot, Scott Kurnit, Steve Maine, Gill Press, Wayne Willis, Bob Worbel, and Les Vadasz.

Special thanks to Renée James, who told me about how Andy Grove called me "Intel's wild duck," hence the name of the book.

The team at Girl Friday Productions deserves special recognition not only for providing all the critical services I needed but for putting up with my micromanagement of the process. This book would not have happened were it not for their comprehensive knowledge and skills in book publishing.

Finally, I want to thank the more than seventy people that I interviewed (mostly on Zoom). They played a critical role in the research that I conducted.

REFERENCES

Much of the work of writing this book was the research I conducted from 2018 through 2021. I reviewed thousands of websites including hundreds of videos. I read over 50 books and interviewed more than 70 key individuals. It was both an enjoyable task and a daunting one as well. Rather than list all the websites and videos here, I have made the links available on this book's website. As a teaser, I have included some of the video links below. Please go to www.WildDuckFlight.com /references to view a more comprehensive list.

WEBSITES AND VIDEOS

THORAXCENTER

Paul Hugenholtz Interview: https://www.youtube.com/watch?v =Tu6hprvtq08

FRANKLIN COMPUTER

Joel Shusterman: https://www.youtube.com/watch?v=RavepbkQP5A
Ace 1100: https://www.youtube.com/watch?v=RYbFY8aljJU

DIGITAL EQUIPMENT CORPORATION

Digital Challenge: Video about the development of the Professional at Digital Equipment: https://www.youtube.com/watch?v =YKbnbvF_2Ew

The Genius of Ken Olsen: https://www.youtube.com/watch?v
=05Hdg7ArQbE&t=5s
Professional 350: https://www.youtube.com/watch?v=OUu3yS9LoMA

DEVELOPMENT OF THE PC

Some History of the IBM PC: https://www.youtube.com/watch?v
=0PceJO3CAGI
Computer History Museum: https://computerhistory.org/blog/personal
-computing-1983-innovation-bursting-in-every-direction/

AVRAM MILLER

Avram testifying about HDTV at Congressional hearing: https://www
.c-span.org/video/?104148-1/digital-high-definition-television
Avram introduces Digital's Personal Computers: https://www.youtube
.com/watch?v=ChVBMfA_P_4
Avram on Smart TV: Computer Chronicles, 1995: https://www.youtube
.com/results?search_query=%22avram+miller%22+computer
+chronicles
Avram Miller Oral History: Computer History Museum, August
2013: https://www.youtube.com/watch?v=L6ktnvRbYAY
Avram NerdTV Interview: https://archive.org/details/nerdtv008
Forbes Interview: https://www.forbes.com/sites/gilpress/2019/06/18
/what-corporate-vcs-can-learn-from-the-first-decade-of-intel
-capital/?sh=63b55c88388a

INTEL

History of MOS Transistor: https://www.youtube.com/watch?v
=q6fBEjf9WPw
Silicon Valley Experience, PBS: https://www.youtube.com/watch?v
=kzit5-Ulx_4
Robert Noyce Interview: https://www.youtube.com/watch?v
=-8RTMFtBjwY
Andy Grove, 1999 Silicon Valley Historical Association: https://www
.youtube.com/watch?v=fvSDUdJHo6k&t=25s

Andy Grove on Strategic Inflection Points: https://www.youtube.com
/watch?v=LfU2Qu4MzZk&t=215s
Remembering Andy Grove: https://www.youtube.com/watch?v
=SQuMoNXMxH0
Oral History Gordon Moore: https://www.youtube.com/watch?v
=gtcLzokagAw
Oral History Les Vadasz: https://ethw.org/Oral-History:Leslie_Vadasz
Oral History Dick Boucher: https://www.youtube.com/watch?v
=0ZFTTGe6j1Y

CABLE BROADBAND

Oral History John Malone: https://www.cablecenter.org/programs
/the-hauser-oral-history-project/m-o-listings/john-malone.html
Oral History Brian Roberts: https://cablecenter.org/the-barco-library
/the-hauser-oral-history-project/r-listings/brian-roberts.html
DOCSIS Development: https://www.youtube.com/watch?v
=9S8jOWwdbO8
DOCSIS Development: John Malone and Richard Green: https://www
.youtube.com/watch?v=Sa43i01qyAY
CEO Round Table on the Information Highway, 1994: https://www
.c-span.org/video/?53677-1/information-highway-ceo
-summit&playEvent

INTERNET

Walter Isaacson: https://www.digitalriptide.org/person/walter
-isaacson/
AOL/Time Warner Merger: https://www.c-span.org/video/?154622-1
/america-online-time-warner-merger
Nathan Myhrvold, 1994: https://www.youtube.com/watch?v
=x8fOqJ2N0Kk

BOOKS

Abbate, Janet. *Inventing the Internet.* Inside Technology. Cambridge,
MA: MIT Press, 1999.

Accenti, Ettore. *My Intel Story: From the Start with Robert Noyce and Gordon Moore till the 50th Anniversary (Marketing Is Fighting)*. Independently published, 2018.

Altraide, Dagogo. *New Thinking: From Einstein to Artificial Intelligence, the Science and Technology That Transformed Our World*. Coral Gables, FL: Mango Publishing, 2019.

Auletta, Ken. *The Highwaymen: Warriors of the Information Superhighway*. San Diego: Harcourt, 1998.

———. *World War 3.0: Microsoft and Its Enemies*. New York, Random House, 2001.

———. *World War 3.0: Microsoft vs. the U.S. Government, and the Battle to Rule the Digital Age*. New York: Broadway Books, 2002.

Band, Jonathan. *Interfaces on Trial: Intellectual Property and Interoperability in the Global Software Industry*. London: Routledge, 2020.

Banks, Michael A. *On the Way to the Web: The Secret History of the Internet and Its Founders*. Apress, 2008.

Berlin, Leslie. *Troublemakers: Silicon Valley's Coming of Age*. New York: Simon & Schuster, 2017.

Bezos, Jeffrey. *Invent and Wander: The Collected Writings of Jeff Bezos*. Boston: Harvard Business Review Press, 2020.

Bloom, John. *Eccentric Orbits: The Iridium Story*. New York: Grove Press, 2016.

Blum, Andrew. *Tubes: A Journey to the Center of the Internet*. New York: Ecco, 2019.

Brilliant, Larry. *Sometimes Brilliant: The Impossible Adventure of a Spiritual Seeker and Visionary Physician Who Helped Conquer the Worst Disease in History*. New York: HarperOne, 2016.

Burgelman, Robert A. *Strategy Is Destiny: How Strategy-Making Shapes a Company's Future*. New York: Free Press, 2002.

Campbell-Kelly, Martin, William Aspray, Nathan Ensmenger, and Jeffrey R. Yost. *Computer: A History of the Information Machine*. 3rd ed. London: Routledge, 2019.

Canion, Rod. *Open: How Compaq Ended IBM's PC Domination and Helped Invent Modern Computing*. Dallas, TX: BenBella Books, Inc, 2013.

Chambers, John, and Diane Brady. *Connecting the Dots: Lessons for Leadership in a Startup World*. New York: Hachette Books, 2018.

Chesbrough, Henry. *Open Innovation Results: Going Beyond the Hype and Getting Down to Business*. Oxford: Oxford University Press, 2020.

Christensen, Clayton M., Taddy Hall, Karen Dillon, and David S. Duncan. *Competing Against Luck: The Story of Innovation and Customer Choice*. New York: Harper Business, 2016.

Christian, Aymar Jean. *Open TV: Innovation beyond Hollywood and the Rise of Web Television*. New York: NYU Press, 2018.

Coll, Steve. *The Deal of the Century: The Breakup of AT&T*. New York: Simon & Schuster, 1986.

Collins, Martin. *A Telephone for the World: Iridium, Motorola, and the Making of a Global Age*. Baltimore, MD: Johns Hopkins University Press, 2018.

Colwell, Robert P. *The Pentium Chronicles: The People, Passion, and Politics behind Intel's Landmark Chips*. Hoboken, NJ: Wiley, 2005.

Coopersmith, Aryae. *Holy Beggars: A Journey from Haight Street to Jerusalem*. El Granada, CA: One World Lights, 2011.

Cortada, James W. *IBM: The Rise and Fall and Reinvention of a Global Icon*. Cambridge, MA: MIT Press, 2019.

Cringely, Robert X. *The Decline and Fall of IBM: End of an American Icon?* NerdTV, 2014.

Daub, Adrian. *What Tech Calls Thinking: An Inquiry into the Intellectual Bedrock of Silicon Valley*. New York: FSG Originals, 2020.

Doerr, John. *Measure What Matters: How Google, Bono, and the Gates Foundation Rock the World with OKRs*. New York: Portfolio/Penguin, 2018.

Fisher, Adam. *Valley of Genius: The Uncensored History of Silicon Valley*. New York: Grand Central Pub, 2018

Gallo, Carmine. *The Innovation Secrets of Steve Jobs: Insanely Different Principles for Breakthrough Success*. New York: McGraw-Hill, 2016.

Garfinkel, Simson L., and Rachel H. Grunspan. *The Computer Book: From the Abacus to Artificial Intelligence, 250 Milestones in the History of Computer Science*. New York: Sterling, 2018.

Gertner, Jon. *The Idea Factory: Bell Labs and the Great Age of American Innovation.* New York: Penguin Press, 2012.

Gilder, George. *Life after Google: The Fall of Big Data and the Rise of the Blockchain Economy.* Washington, DC: Gateway Editions, 2018.

Gillett, Stephen. *From Simi Valley to Silicon Valley: A Story of Hard Work, Serendipity, and Questing.* Austin, TX: Lioncrest Publishing, 2019.

Grove, Andrew S. *High Output Management.* 2nd ed. New York: Vintage, 1995.

———. *Only the Paranoid Survive: How to Exploit the Crisis Points That Challenge Every Company and Career.* London: Profile Books, 1998.

Hafner, Katie, and Matthew Lyon. *Where Wizards Stay Up Late: The Origins of the Internet.* New York: Touchstone, 1996.

Hagey, Keach. *The King of Content: Sumner Redstone's Battle for Viacom, CBS, and Everlasting Control of His Media Empire.* New York: Harper Business, 2018.

Hertzfeld, Andy. *Revolution in the Valley: The Insanely Great Story of How the Mac Was Made.* Sebastopol, CA: O'Reilly Media, 2011.

Hey, Tony, and Gyuri Pápay. *The Computing Universe: A Journey through a Revolution.* New York: Cambridge University Press, 2015.

Hiltzik, Michael A. *Dealers of Lightning: Xerox PARC and the Dawn of the Computer Age.* New York: Harper Business, 2000.

Horowitz, Ben. *The Hard Thing about Hard Things: Building a Business When There Are No Easy Answers.* New York: Harper Business, 2014.

Isaacson, Walter. *The Innovators: How a Group of Inventors, Hackers, Geniuses, and Geeks Created the Digital Revolution.* New York: Simon & Schuster, 2014.

Janeway, William H. *Doing Capitalism in the Innovation Economy: Reconfiguring the Three-Player Game between Markets, Speculators and the State.* 2nd ed. Cambridge, UK: Cambridge University Press, 2018.

Kahney, Leander. *Tim Cook: The Genius Who Took Apple to the Next Level.* New York: Portfolio/Penguin, 2019.

Kidder, Tracy. *The Soul of a New Machine.* New York: Modern Library, 1997.

Klein, Alec. *Stealing Time: Steve Case, Jerry Levin, and the Collapse of AOL Time Warner.* New York: Simon & Schuster, 2004.

Knowling, Bob. *You Can Get There from Here: My Journey from Struggle to Success.* New York: Portfolio/Penguin, 2011.

Kupor, Scott. *Secrets of Sand Hill Road: Venture Capital and How to Get It.* New York: Portfolio/Penguin, 2019.

Kushner, David. *The Players Ball: A Genius, a Con Man, and the Secret History of the Internet's Rise.* New York: Simon & Schuster, 2019.

Lashinsky, Adam. *Inside Apple: How America's Most Admired—and Secretive—Company Really Works.* New York: Business Plus, 2013.

Lécuyer, Christophe. *Making Silicon Valley: Innovation and the Growth of High Tech, 1930–1970.* Cambridge, MA: MIT Press, 2006.

Levy, Lawrence. *To Pixar and Beyond: My Unlikely Journey with Steve Jobs to Make Entertainment History.* Boston: Mariner Books, 2016.

Levy, Steven. *Facebook: The Inside Story.* New York: Blue Rider Press, 2020.

———. *Hackers: Heroes of the Computer Revolution.* Sebastopol, CA: O'Reilly Media, 2010.

Lewis, Michael. *The New New Thing: A Silicon Valley Story.* New York: W. W. Norton, 2014.

Livingston, Jessica. *Founders at Work: Stories of Startups' Early Days.* New York: Apress, 2008.

Lockman, Brian, and Don Sarvey. *Pioneers of Cable Television: The Pennsylvania Founders of an Industry.* Jefferson, NC: McFarland & Company, Inc, 2005.

Lotz, Amanda D. *The Television Will Be Revolutionized.* 2nd ed. New York: NYU Press, 2014.

———. *We Now Disrupt This Broadcast: How Cable Transformed Television and the Internet Revolutionized It All.* Cambridge, MA: MIT Press, 2018.

Malik, Om. *Broadbandits: Inside the $750 Billion Telecom Heist.* Hoboken, NJ: Wiley, 2003.

Malone, Michael S. *The Intel Trinity: How Robert Noyce, Gordon Moore, and Andy Grove Built the World's Most Important Company.* New York: Harper Business, 2014.

Markoff, John. *What the Dormouse Said: How the Sixties Counterculture Shaped the Personal Computer Industry.* New York: Penguin Books, 2006.

Marshall, Stephen J. *The Story of the Computer: A Technical and Business History.* CreateSpace, 2017.

Mayer-Schönberger, Viktor, and Kenneth Cukier. *Big Data: The Essential Guide to Work, Life and Learning in the Age of Insight.* New and expanded edition. London: John Murray, 2013.

McCullough, Brian. *How the Internet Happened: From Netscape to the iPhone.* Liveright, 2018.

McNamee, Roger. *Zucked: Waking up to the Facebook Catastrophe.* First ed. London: HarperCollins, 2019.

Menuez, Doug. *Fearless Genius: The Digital Revolution in Silicon Valley, 1985–2000.* New York: Atria Books, 2014.

Messina, Michelle E., and Jonathan C. Baer. *Decoding Silicon Valley: The Insider's Guide.* Redwood City, CA: Decode Publishers, 2016.

Mullen, Megan. *Television in the Multichannel Age: A Brief History of Cable Television.* Malden, MA: Wiley-Blackwell, 2008.

Munk, Nina. *Fools Rush in: Steve Case, Jerry Levin, and the Unmaking of AOL Time Warner.* New York: Harper Business, 2009.

Nicholas, Tom. *VC: An American History.* Cambridge, MA: Harvard University Press, 2019.

Nicholson, Matt. *When Computing Got Personal: A History of the Desktop Computer.* Bristol, UK: Matt Publishing, 2014.

Ovitz, Michael. *Who Is Michael Ovitz?* London: WH Allen, 2019.

Parsons, Patrick R. *Blue Skies: A History of Cable Television.* Philadelphia: Temple University Press, 2008.

Paternot, Stephan. *A Very Public Offering: The Story of theglobe.com and the First Internet Revolution.* New York: Actarus Press, 2018.

Perkins, Thomas. *Valley Boy: The Education of Tom Perkins.* New York: Gotham, 2007.

Randolph, Marc. *That Will Never Work: The Birth of Netflix and the Amazing Life of an Idea.* Little, Brown and Company, 2019.

Rao, Arun, and Piero Scaruffi. *A History of Silicon Valley: The Greatest Creation of Wealth in the History of the Planet*. Palo Alto, CA: Omniware, 2011.

Reid, T. R. *The Chip: How Two Americans Invented the Microchip and Launched a Revolution*. New York: Random House, 2001.

Robichaux, Mark. *Cable Cowboy: John Malone and the Rise of the Modern Cable Business*. Hoboken, NJ: Wiley, 2005.

Ryan, Johnny. *A History of the Internet and the Digital Future*. London: Reaktion Books, 2010.

Satkowiak, Larry. *The Cable Industry: A Short History through Three Generations*. Denver, CO: The Cable Center, 2015.

Schein, Edgar H. *DEC Is Dead, Long Live DEC: The Lasting Legacy of Digital Equipment Corporation*. San Francisco, CA: Berrett-Koehler Publishers, 2003.

Schlender, Brent, and Rick Tetzeli. *Becoming Steve Jobs: The Evolution of a Reckless Upstart into a Visionary Leader*. Toronto: Signal, 2016.

Schmidt, Eric, Jonathan Rosenberg, and Alan Eagle. *Trillion Dollar Coach: The Leadership Playbook of Silicon Valley's Bill Campbell*. New York: Harper Business, 2019.

Shah, Tarang, and Shital Shah. *Venture Capitalists at Work-German: How VCs Identify and Build Billion-Dollar Successes*. Apress, 2013.

Silver, Nate. *The Signal and the Noise: Why So Many Predictions Fail—but Some Don't*. New York: Penguin Books, 2015.

Simpson, Edward N. *TV History: The Most Important Technologies, People and Events That Created the History of Television*. 2019.

Soni, Jimmy, and Rob Goodman. *A Mind at Play: How Claude Shannon Invented the Information Age*. New York: Simon & Schuster, 2017.

Stross, Randall E. *eBoys: The First Inside Account of Venture Capitalists at Work*. New York: Ballantine Books, 2001.

Summary: Andy Grove: Review and Analysis of Tedlow's Book. United States: Business Book Summaries, 2013.

Swisher, Kara. *There Must Be a Pony in Here Somewhere: The AOL Time Warner Debacle and the Quest for a Digital Future*. New York: Crown Business, 2004.

Taplin, Jonathan. *Move Fast and Break Things: How Facebook, Google, and Amazon Cornered Culture and Undermined Democracy.* New York: Back Bay Books, 2018.

Tedlow, Richard S. *Giants of Enterprise: Seven Business Innovators and the Empires They Built.* New York: HarperBusiness, 2003.

Waldrop, M. Mitchell. *The Dream Machine.* 4th ed. San Francisco: Stripe Press, 2018.

Wheeler, Tom. *From Gutenberg to Google: The History of Our Future.* Washington, DC: Brookings Institution Press, 2019.

Wiedeman, Reeves. *Billion Dollar Loser: The Epic Rise and Spectacular Fall of Adam Neumann and WeWork.* New York: Little, Brown and Company, 2020.

Williams, Christopher. *The Battle for Sky: The Murdochs, Disney, Comcast, and the Future of Entertainment.* Bloomsbury Business, 2019.

Wolff, Michael. *Burn Rate: How I Survived the Gold Rush Years on the Internet.* New York: Simon & Schuster, 1999.

———. *The Man Who Owns the News: Inside the Secret World of Rupert Murdoch.* New York: Broadway Books, 2008.

Wolk, Alan. *Over the Top: How the Internet Is (Slowly but Surely) Changing the Television Industry.* CreateSpace, 2015.

Yoffie, David B., and Michael A. Cusumano. *Competing on Internet Time: Lessons from Netscape and Its Battle with Microsoft.* New York: Free Press, 1998.

———. *Strategy Rules: Five Timeless Lessons from Bill Gates, Andy Grove, and Steve Jobs.* New York: Harper Business, 2015.

Zuckerman, Gregory. *The Man Who Solved the Market: How Jim Simons Launched the Quant Revolution.* New York: Penguin, 2019.

MARLON BRANDO

March 27, 1997

Mr. Sriram Viswanathan
Intel Corporation
9830 Wilshire Boulevard
Beverly Hills, CA 90212

Dear Sriram:

Yesterday I was engulfed. I got sucked into the black hole of technology. I was dazed; I found myself in intellectual free-fall. I think, for the first time, I had a tangible sense of the seismic wave of change that is going to forever and completely alter our lives, our belief systems, our social values and moral prospectives, in fact, everything. I don't think that we are, as a species, psychologically prepared for this extraordinary evolution of technology. We have almost always misused and misapplied technology with disastrous results. It would seem now, at this juncture, the fate of this species is in a race between its animalistic nature and the salvation that wisely applied genetic mutation will bring.

In any event, yesterday's experience, like a bowling ball, hit many intellectual and psychological pins in such a random fashion that even at this writing I can hardly make sense of it except that it's plain to me now, as never before, the world, as it is perceived, is gone and no one, or let me say, few, if any, can have any sense of what the human destiny holds in store. I often feel as though we are on a magnetic levitated train going 700 miles an hour not knowing how much track is in front of us while we are trying to invent a braking system. What an afternoon.

Thank you all for the experience of yesterday. I would like very much to stay in touch because I would like to know when we reach the falls.

Warm regards,

Marlon Brando

MB/ceb

Marlon Brando gets it.

ABOUT THE AUTHOR

Avram Miller is an American-born businessperson, venture capitalist, scientist, technologist, and musician. He is best known for his work at Intel, where he served as vice president, business development. Together with Les Vadasz, he cofounded Intel Capital and led Intel's successful initiative to create residential broadband.

Miller's leadership in developing both the technology and business infrastructure for residential broadband laid an important foundation for the construction of today's internet. In 1996, *USA Today* called Miller "a one-man think tank." In the same article, Brian Roberts, CEO of Comcast, gave Miller "much of the credit" for the development of the cable modem.

Miller's work in venture capital is well recognized. In 2003, Miller was number eight on the *Forbes* Midas List of the top one hundred people in venture capital. While at Intel, he managed a multibillion-dollar portfolio, which included early investments in Broadcast.com, CNET, Verisign, and Covad.

After leaving Intel in 1999, Miller founded The Avram Miller Company, providing strategic advice to technology companies throughout the world. He has served as a senior advisor to Lazard, sat on the boards of many public and private internet companies, and served on the boards of entertainment companies, including Maxis and King World Productions. Currently, he is on the advisory board of Sommetrics, a sleep-health company.

Miller has been active in nonprofit work. He was the founding chair of PluggedIn, a senior advisor to Equal Access, and a trustee of the California Institute of the Arts. He currently splits his time between Israel and the United States and enjoys traveling with his wife.

Printed in Great Britain
by Amazon